Biologic Foundations for Skeletal Tissue Engineering

Biologic Foundations for Skeletal Tissue Engineering
Ericka M. Bueno and Julie Glowacki

ISBN: 978-3-031-01455-0   paperback
ISBN: 978-3-031-02583-9   ebook

DOI: 10.1007/978-3-031-012583-9

A Publication in the Springer series
SYNTHESIS LECTURES ON TISSUE ENGINEERING

Lecture #7
Series Editor: Kyriacos A. Athanasiou and J. Kent Leach, *University of California, Davis*
Series ISSN
Synthesis Lectures on Tissue Engineering
Print 1944-0316   Electronic 1944-0308

Synthesis Lectures on Tissue Engineering

Editor

Kyriacos A. Athanasiou and J. Kent Leach, *University of California, Davis*

The Synthesis Lectures on Tissue Engineering series will publish concise books on aspects of a field that holds so much promise for providing solutions to some of the most difficult problems of tissue repair, healing, and regeneration. The field of Tissue Engineering straddles biology, medicine, and engineering, and it is this multi-disciplinary nature that is bound to revolutionize treatments for a plethora of tissue and organ problems. Central to Tissue Engineering is the use of living cells with a variety of biochemical or biophysical stimuli to alter or maximize cellular functions and responses. However, in addition to its therapeutic potentials, this field is making significant strides in providing diagnostic tools. Each book in the Series will be a self-contained treatise on one subject, authored by leading experts. Books will be approximately 65-125 pages. Topics will include 1) Tissue Engineering knowledge on particular tissues or organs (e.g., articular cartilage, liver, cardiovascular tissue), but also on 2) methodologies and protocols, as well as 3) the main actors in Tissue Engineering paradigms, such as cells, biomolecules, biomaterials, biomechanics, and engineering design. This Series is intended to be the first comprehensive series of books in this exciting area.

Biologic Foundations for Skeletal Tissue Engineering
Ericka M. Bueno and Julie Glowacki
2011

Mona K. Marei
2010

Cells and Biomaterials for Intervertebral Disc Regeneration
Sibylle Grad, Mauro Alini, David Eglin, Daisuke Sakai, Joji Mochida, Sunil Mahor, Estelle Collin, Biraja Dash, Abhay Pandit
2010

Fundamental Biomechanics in Bone Tissue Engineering
X. Wang, J.S. Nyman, X. Dong, H. Leng, M. Reyes
2010

Articular Cartilage Tissue Engineering
Kyriacos A. Athanasiou, Eric M. Darling, Jerry C. Hu
2009

iv

Tissue Engineering of Temporomandibular Joint Cartilage
Kyriacos A. Athanasiou, Alejandro J. Almarza, Michael S. Detamore, Kerem N. Kalpakci
2009

Engineering the Knee Meniscus
Kyriacos A. Athanasiou, Johannah Sanchez-Adams
2009

Biologic Foundations for Skeletal Tissue Engineering

Ericka M. Bueno
Brigham and Women's Hospital
Harvard Medical School

Julie Glowacki
Brigham and Women's Hospital
Harvard Medical School
Harvard School of Dental Medicine

SYNTHESIS LECTURES ON TISSUE ENGINEERING #7

ABSTRACT

Tissue engineering research for bone and joint applications entails multidisciplinary teams bringing together the needed expertise in anatomy, biology, biochemistry, pathophysiology, materials science, biomechanics, fluidics, and clinical and veterinary orthopedics. It is the goal of this volume to provide students and investigators who are entering this exciting area with an understanding of the biologic foundations necessary to appreciate the problems in bone and cartilage that may benefit from innovative tissue engineering approaches. This volume includes state-of-the-art information about bone and cartilage physiology at the levels of cell and molecular biology, tissue structure, developmental processes, their metabolic and structural functions, responses to injury, mechanisms of post-natal healing and graft incorporation, the many congenital and acquired disorders, effects of aging, and current clinical standards of care. It reviews the strengths and limitations of various experimental animal models, sources of cells, composition and design of scaffolds, activities of growth factors and genes to enhance histogenesis, and the need for new materials in the context of cell-based and cell-free tissue engineering. These building blocks constitute the dynamic environments in which innovative approaches are needed for addressing debilitating disorders of the skeleton. It is likely that a single tactic will not be sufficient for different applications because of variations in the systemic and local environments. The realizations that tissue regeneration is complex and dynamic underscore the continuing need for innovative multidisciplinary investigations, with an eye to simple and safe therapies for disabled patients.

KEYWORDS

bone, cartilage, structure, function, properties, aging, development, modeling, remodeling, fracture healing, bone pathology, cartilage pathology, animal models, scaffolds, cell-based tissue engineering, cell-free tissue engineering, growth factors, differentiation factors

Contents

Acknowledgments ix

1   Introduction 1

2   Structure and Function of Bone and Cartilage Tissue 3
     2.1   Bone 3
         2.1.1   Structure 3
         2.1.2   Function 16
         2.1.3   Mechanical Properties 21
     2.2   Articular Cartilage 30
         2.2.1   Structure 30
         2.2.2   Function 38
         2.2.3   Mechanical Properties 39

3   Development 45
     3.1   Intramembranous Bone Formation 45
     3.2   Endochondral Bone Formation 45
     3.3   Articular Cartilage 51

4   Responses to Injury and Grafting 53
     4.1   Bone 53
         4.1.1   Microfractures 53
         4.1.2   Bone Fractures 53
         4.1.3   Fracture Healing 54
         4.1.4   Distraction Osteogenesis 56
         4.1.5   Comparison of Post-natal Fracture Healing and Skeletal Embryogenesis   56
         4.1.6   Bone Grafts, Implants, and Bone Substitute Materials 61
     4.2   Cartilage 64

5   Clinical Applications for Skeletal Tissue Engineering 67
     5.1   Disorders in Bone 67
         5.1.1   Focal Deficiencies 67

          5.1.2   Systemic Disorders of Bone 71

    5.2   Cartilage 76

          5.2.1   Osteoarthritis 76

          5.2.2   Rheumatoid Arthritis 79

          5.2.3   Osteochondritis Dissecans 81

          5.2.4   Temporomandibular Joint (TMJ) Disorders 82

6   **Animal Models** **85**

    6.1   Bone 85

          6.1.1   Biocompatibility 85

          6.1.2   Osteocompatibility 86

          6.1.3   Bone Substitutes 86

          6.1.4   Clinical Relevance 90

    6.2   Cartilage 93

7   **Tissue Engineering Principles for Bone and Cartilage** **95**

    7.1   Bone 95

          7.1.1   Scaffolds 96

          7.1.2   Bioactive Factors 101

          7.1.3   Cells 104

          7.1.4   *In Vitro* Studies 106

          7.1.5   *In Vivo* Studies 107

    7.2   Cartilage 109

          7.2.1   Scaffolds 110

          7.2.2   Bioactive Factors 114

          7.2.3   Cells 116

          7.2.4   *In Vitro* Studies 118

          7.2.5   *In Vivo* Studies 120

8   **Perspectives** **123**

    References **127**

    Authors' Biographies **209**

Acknowledgments

The authors appreciate the assistance with the figures from Karen Aneshansley. With many thanks,

Ericka M. Bueno and Julie Glowacki
January 2011

CHAPTER 1

Introduction

Bone is an organized, highly complex tissue. It provides both mechanical and metabolic functions. The mechanical ones include structure and support of an organism, leverage for musculoskeletal locomotion, and protection of vital organs such as the brain and heart. The metabolic ones involve contributions to mineral homeostasis, wherein calcium phosphate mineral is deposited during times of plenty and can be resorbed during times of need. Its mechanical strength and elasticity allow it to resist large forces during those normal functions. The abundant vascularity of bone and cell-mediated remodeling activities provide it with a remarkable capacity for true regeneration without scarring. Nevertheless, some osseous defects do not fully heal in the absence of a grafting material. Traditional therapies for large or non-union bone defects involve the use of autogenous or allogeneic bone grafts; however, their shortcomings motivate research on novel therapies for bone regeneration [1, 2]. In order to develop new solutions to skeletal problems, a thorough understanding of bone's anatomy, physiology, mechanical properties, and response to injury is necessary.

Articular cartilage covers the ends of bones in diarthrodial joints. It is a dense, relatively hypocellular tissue that provides joints with an almost frictionless articulating surface, with lubrication, and with the ability to absorb mechanical load and transfer it to the distal bone. Chronic wear, trauma, or degenerative joint diseases can severely and irreversibly damage articular cartilage. Unlike bone, cartilage has a limited capacity for self-repair; thus, damage to articular cartilage eventually leads to severe and progressive disability, with degeneration of the articular surface often to the point of complete destruction [3]. Available treatments may be palliative surgical approaches that are relatively simple such as removal and debridement of the damaged areas in order to reduce mechanical irritation and inflammation, or they may be reparative and promote the formation of reactive tissue, such as in the case of microfracture. More complex strategies aim to restore the articular cartilage surface with, for example, autologous chondrocyte implantation or osteochondral grafting [4]. These approaches may provide relief even if they generate fibrous tissue rather then continuous neocartilage [5]. Surface reconstruction, hemiarthoplasty, or total joint replacement may be required for extensive loss of articular cartilage. Over the last two decades, considerable advances in the understanding of articular cartilage biology, physiology, and biomechanics have resulted in biologically-based alternatives to current treatments, with the objective of a viable, tissue-engineered cartilage replacement.

CHAPTER 2

Structure and Function of Bone and Cartilage Tissue

2.1 BONE

2.1.1 STRUCTURE

2.1.1.1 Tissue Composition

Bone tissue is composed of several types of cells and a biphasic extracellular matrix (ECM). The ECM of bone is approximately 30% an organic phase and 70% inorganic mineral by weight. The organic phase comprises mainly (98%) collagen type I fibers that provide elasticity and structure. The collagen fibers generally align parallel to the long axis of bone [6]. Also present in the organic phase in smaller proportions are other structural and biologically active proteins, proteoglycans, glycoproteins, peptides, lipid materials, and adsorbed serum proteins. These constituents provide structural properties and/or participate in signaling processes related to matrix organization, mineralization, and turnover.

The inorganic portion of bone ECM is composed of calcium phosphate mineral crystals and provides strength and rigidity. The minerals are deposited as a poorly-crystalline carbonate-apatite structure similar to, but less ordered than calcium and phosphate-containing geologic hydroxyapatite [7]. Mineral maturation to more ordered crystal structure occurs thermodynamically through the rapid formation of crystals from the saturated body fluid, and renders the inorganic phase into a more crystalline apatitic structure, containing (on a weight basis, in human adults) phosphorous (P, 15.2%), carbonate (CO_3^{2-}, 7.4%), sodium (Na^+, 0.9%), magnesium (Mg^{2+}, 0.72%), chloride (Cl^-, 0.13%), pyrophosphate ($P_2O_7^{4-}$, 0.07%), potassium (K^+, 0.03%), fluoride (F^-, 0.03%), and other trace elements [8]. The mineral crystals are small, with an approximate diameter of 25 to 75Å and approximate length of 200Å. These dimensions provide the mineral phase of bone with an enormous surface area that has been estimated as approximately 100 m^2 per gram [6]. The small mineral crystals constitute the resorbable reservoir of mineral ions for calcium and phosphate homeostasis. The amount of mineral per unit volume, termed bone mineral density (BMD), contributes to the stiffness of the tissue.

2.1.1.2 Cells

Bone has three major cell types. Osteoblasts are responsible for bone formation and are found at the surface of bone; a fraction of them become mature osteocytes embedded within the ECM. Osteoclasts are the large multinucleated cells that are responsible for bone resorption (Figure 2.1).

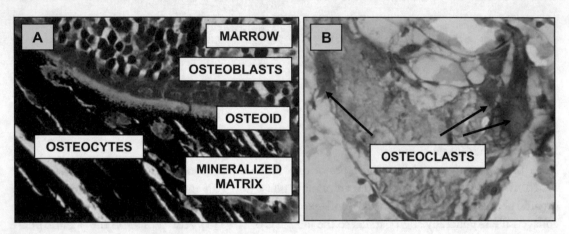

Figure 2.1: (a) Histological section of bone shows hematopoietic marrow, layer of plump osteoblasts, pink-staining, non-mineralized osteoid matrix, black-staining mineralized matrix, and osteocytes in lacunae within the matrix (von-Kossa/H&E stain). (b) Histological section of bone shows red-staining, multinucleated osteoclasts on the surfaces of a blue-staining bone trabeculum (stained red for tartrate-resistant acid phosphatase activity and counterstained with hematoxylin).

Osteoblasts are cuboidal in shape, and as highly secretory cells, they contain abundant rough endoplasmic reticulum. In growing bone, they are found at a distance from the mineralized matrix of bone in a zone of unmineralized bone matrix, termed osteoid. In addition to their role in the synthesis and secretion of organic matrix, osteoblasts produce paracrine and autocrine factors that result in recruitment of osteoprogenitor cells, growth of preosteoblasts, and regulation of osteoclastic resorption of the mineralized bone matrix [9].

Osteoblasts are derived from stem/progenitor cells that have the potential to differentiate into other types of cells, such as chondrocytes, adipocytes, and fibroblasts. These multipotential mesenchymal progenitor cells have been called mesenchymal stem cells, or marrow-derived stromal cells, when isolated from the adherent fraction of adult bone marrow [10]. They have also been isolated from diverse tissues such as adipose tissue [11], muscle [12], synovium [13], dental pulp [14], periodontal ligament [15], amniotic fluid [16], umbilical cord and umbilical cord blood [17], and peripheral blood [18].

Osteocytes are mature, post-synthetic osteoblasts that have become encased in the mineralized matrix. Only a fraction of osteoblasts become osteocytes, the majority undergoing apoptosis after they secrete the bone ECM [19]. Compared with osteoblasts, the quiescent osteocytes have fewer synthetic organelles and a higher nucleus-to-cytoplasm ratio. Osteocytes are found within the Haversian system of lamellar bones, in lacunae communicating with Haversian canals from which they derive nutrients and oxygen, and, in general, are no more than 0.1 mm away from a capillary vessel [6, 20]. Osteocytes are interconnected through thin cytoplasmic processes in canaliculi, creating a large intercellular network that enables them to engage in mineral homeostasis, mechanical sensing, and signal transduction (Figure 2.2).

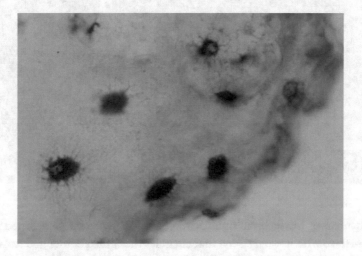

Figure 2.2: Histological section of bone shows blue-staining matrix and purple osteocytes with interconnecting cytoplasmic processes (Toluidine blue stain).

The bone-resorbing cells, osteoclasts, originate from the hematopoietic lineage stem cells in the non-adherent portion of marrow. Osteoclasts are large (20 to 100 μm in diameter) and multi-nucleated, forming from the fusion of mononuclear precursor cells. Osteoclasts attach to the bone surface, isolating areas of tissue termed Howship's lacunae or "pits of bone resorption," into which they secrete hydrogen ions that lower the pH and dissolve the calcium phosphate mineral. Demineralization renders the organic matrix susceptible to acidic proteolytic digestion. The products of matrix degradation are incorporated into coated vesicles and vacuoles for translocation and elimination. Active osteoclasts in pits of bone resorption exhibit a characteristic ruffled border opposed to the surface of bone (Figure 2.3). In cancellous bone, osteoclasts are found in Howship lacunae or erosions on the surface of trabeculae. In human cortical bone, osteoclasts are often located at the advancing end of the cutting cones, the tunnels in which the mineral and organic matrix of bone are resorbed (Figure 2.4). Osteoclasts are closely associated with small blood vessels and at a distance from bone-forming osteoblasts that fill the void left by the cutting cone. The osteoblasts synthesize concentric rings of lamellar bone around the central blood vessel and produce a new osteon. The action of the cutting cones results in overlapping osteons and the renewal of bone matrix. This

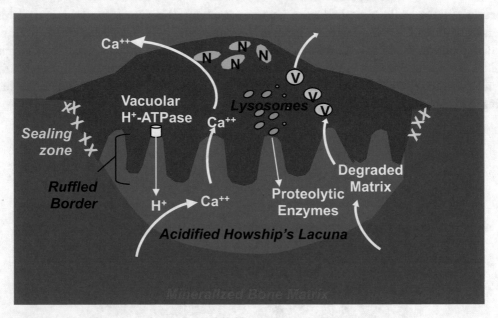

Figure 2.3: Mechanism of Osteoclastic Bone Resorption. An active multinucleated (N) osteoclast adheres to the bone surface with a sealing zone that encircles the ruffled border and isolates the subcellular Howship's lacuna. Secretion of hydrogen ions with a vacuolar hydrogen-ATPase creates a low pH environment, in which calcium ions dissolve from the matrix and are pumped into the circulation. Demineralization renders the organic matrix susceptible to proteolysis by acidic enzymes such as cathepsin K, matrix metalloproteinase 9, and lysosomal enzymes. Large coated vacuoles (V) transport the degradation products for elimination.

process of destruction of former generations of osteons and their replacement with a new generation continues throughout life. Thus, bone-forming osteoblasts and bone-resorbing osteoclasts interact in a precisely regulated manner to achieve skeletal growth and to maintain skeletal remodeling. Imbalances in their activities are features of various pathological conditions, such as osteopetrosis, osteoporosis, and other osteolytic and osteosclerotic bone diseases.

2.1.1.3 Extracellular Matrix (ECM)

Collagen type I is the major protein in the organic ECM of bone. The collagen molecule is composed of three chains, each being a left-handed helix with three amino acid residues per turn. One-third of the amino acids in collagen are glycine with the repeat sequence of Glycine-X-Y, with high content of proline and hydroxyproline as X and Y. Each chain is a long rod with just over 1000 amino acids in the triplet array and short non-triplet regions at each terminus. The type I collagen molecule has two identical chains called α1, and one chain with a different amino acid sequence, termed α2. Collagen is

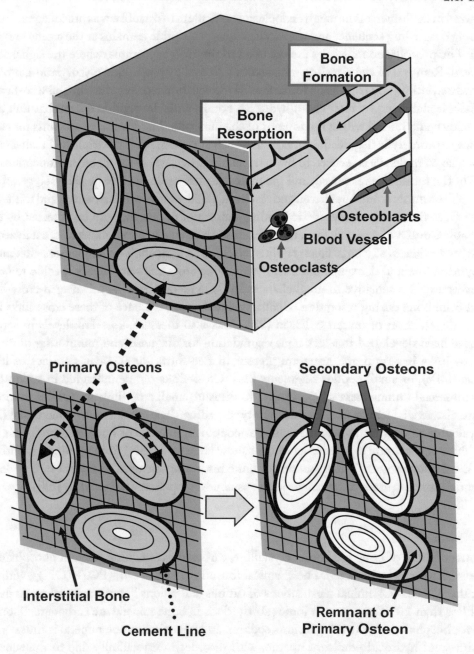

Figure 2.4: The internal remodeling of human cortical bone. At the microscopic level, cortical bone is composed of tubes of organized osteons, each with a central vascular channel and bounded by a dense "cement" line, within a more random interstitial matrix. Numerous cutting cones course through the bone, replacing old bone and generating new osteons, with remnants of previous osteons.

synthesized in the lumen of the rough endoplasmic reticulum of osteoblasts as procollagen, a molecule containing a signaling sequence and non-collagenous propeptide domains at the amino and carboxy termini. The procollagen molecules are extruded to the Golgi apparatus, where the signal sequence is removed. Removal of the signaling sequence is followed by propeptide glycosylation, proline and lysine hydroxylation, and triple helix formation. Hydroxylation requires vitamin C as a co-factor. The triple helix is stabilized by inter-chain hydrogen bonds, with the peptide bonds that link adjacent amino acids being buried within the interior of the molecule. This geometry accounts for collagen's resistance to proteolytic degradation by enzymes such as pepsin. After the triple helix self-assembles, the procollagen molecule is secreted to the extracellular space, where the globular propeptides at both termini of the molecule are cleaved, and the molecules self-assemble into quarter-staggered fibrillar arrays. Self-assembly is entropy-driven and results in a side-to-side overlap pattern and a tail-to-head gap that gives rise to the characteristic banding pattern of 67 nm that can be detected by electron microscopy. Covalent cross-link formation occurs at this stage between molecules, catalyzed by the enzyme lysyl oxidase [21]; cross links stabilize the collagen network. Two major cross-link molecules exist, pyridinoline and deoxypyridinoline. Pyridinoline cross-links are highly specific to bone and their levels are very sensitive to small changes in bone metabolism [22]. Collagen cross-links are released from bone during resorption or collagen breakdown; presence of these cross-links in urine indicates degradation of mature collagen [23]. Assays to measure cross-link levels in serum and urine have been developed to assess bone degradation for diagnosis and monitoring of treatment. The cross-links in serum and urine are present in free-form (40%), small peptides of less than 1000 Da (30%), or small peptide fragments [24]. Cross-links can be measured in serum or urine with commercial immunoassays [25]. Assays to measure small cross-linked peptides in serum and urine are also available, but there is uncertainty regarding their specificity to bone tissue [26, 27]. To be valuable in diagnosing and managing osteoporosis, markers of bone resorption in a patient should show consistently repeatable levels (not show large variability) from month to month, day to day, or even hour to hour. These markers of bone resorption are used to predict rate of bone loss, fracture risk, and response to treatment in the osteoporotic patient (reviewed by [28]).

2.1.1.4   Mineral

The mineral phase of bone is a poorly crystalline, carbonate-containing calcium phosphate analog of geologic hydroxyapatite (HA). The chemical formula of HA is $Ca_{10}(PO_4)_6(OH)_2$, with a Ca/P molar ratio of 1.7:1. Chemical substitutions of cations and anions in bone mineral result in a Ca/P ratio of less than 1.7. The common ionic substitutions in bone mineral are carbonate, fluoride, citrate, pyrophosphate, chloride, magnesium, sodium, and potassium. Bone mineral is thus considered a calcium- and hydroxide-deficient apatite, with disorder in crystallinity due to the substituting components.

Mineralization of the ECM of bone begins in multiple independent nucleation sites in the hole regions between the heads and tails of collagen that result from the lateral quarter-stagger arrangement of the collagen molecules [29] (Figure 2.5). The normal concentration of ionized

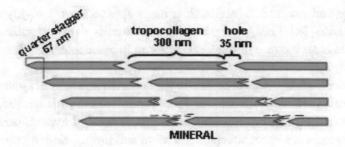

Figure 2.5: The lateral quarter-stagger arrangement of collagen molecules in a collagen fiber shows the hole zones and their roles as independent nucleating sites in bone mineral formation.

(free) calcium in most species is 1.25 mM. The concentration of inorganic phosphate is variable, fluctuating with dietary intake, but the fasting level is 1 mM. Thus, the degree of saturation of body fluids can be expressed as a simple ion product $[Ca^{2+}] \times [P_{inorganic}]$, or 1.3 mM $\times$ 1 mM = 1.3 mM^2. That value is metastable, or close to the solubility of hydroxyapatite. Mineral crystals will precipitate from such an environment once the thermodynamic barrier is overcome by either a nucleation site or by a local increase in either ion. Matrix components such as phosphoproteins may facilitate nucleation in the hole regions in the collagen matrix [30]. After nucleation, crystal growth is thermodynamically favored at the physiological concentrations of ions in the body fluid. To further facilitate crystal growth, alkaline phosphatase, an enzyme secreted by the osteoblast, acts to raise the local concentration of inorganic phosphates to levels above saturation by hydrolysis of organic phosphates [31]. Crystal growth involves the incorporation of the other ions plentiful in the environment, especially carbonate and sodium.

Once calcium-carbonato-phosphate nucleation occurs, crystal growth proceeds thermodynamically in proportion to the concentrations of the mineral components [32]. This raises a problem similar to the tale of Lot's wife and what it is that prevents the complete mineralization of all collagen-containing tissues. This problem is solved by an array of natural small and large molecular weight inhibitors. The first inhibitor to be identified was pyrophosphate, which at physiological concentrations in plasma, urine, and saliva inhibits most forms of calcification. Mineralization will not proceed until the inhibitor is inactivated. Alkaline phosphatase, among other enzymes, hydrolyzes pyrophosphate to inactive orthophosphate. Other natural inhibitors include ions such as citrate, magnesium, and large molecular weight molecules such as glycosaminoglycans.

The initial mineral crystal size is small, $20 \times 5 \times 5$ nm, a size that accommodates bone mineral solubility and exchange. The nano-crystallinity of bone mineral results in high surface area (estimated at 50 to 250 m^2/g) and solubility and is at the heart of the metabolic function of bone to serve as a mobilizable reservoir of exchangeable mineral. In addition, carbonate-containing apatite has higher solubility than carbonate-free apatite [33]. Investigations of composition of bone mineral and its x-ray diffraction characteristics indicate that the mineral phase of bone becomes more crystalline with age and maturation, although it never approaches the highly crystalline state of geologic hydroxyapatite [34]. Thus, bone mineral is deposited as poorly crystalline carbonatoapatite, and with time, there is an increase in the Ca/P ratio, an increase in crystal size, and an increase in crystallinity [35].

Although there is an age-related decline in the volumetric density of bone as it becomes more porous [36], the bone matrix becomes more densely mineralized with age [35, 37]. When there is a larger proportion of older, more mineralized matrix, it is called hypermineralization [38]. The mechanical consequences are increased crack initiation and propagation, increased brittleness, and increased fragility. Thus, there is a range of mineralization within which bone's modulus of toughness is maximized.

Elegant studies used Fourier transformed infrared (FTIR) microspectroscopy to observe mineral composition and crystallinity within the osteonal structure of bone. The results provided striking evidence of maturation from the area of most recent mineral deposition adjacent to the Haversian canal to the oldest mineral on the periphery of the osteon [39]. The study showed a decrease in the carbonate-to-phosphate ratio and an increase in crystal size and order from the center to the periphery of an osteon. This thermodynamic transformation with time decreases the solubility of the mineral phase, a phenomenon that could have untoward consequences for mineral homeostasis if it were to continue unabated. The normal turnover of bone's organic and mineral matrix that is achieved by the coordination of osteoclastic resorption and osteoblastic bone formation in bone remodeling can be viewed, in part, as a process of mineral rejuvenation that is central for mineral exchange and homeostasis. Unfortunately, this process is not perfectly balanced.

The importance of bone remodeling in controlling mineral crystallinity is best demonstrated by studies in fish species that lack bone remodeling. In many teleosts, bone is produced by osteoblasts without their being incorporated into the matrix; thus, the bone tissue lacks osteocytes and is called acellular bone [40]. It is striking that acellular (anosteocytic) and cellular bone can be present within a single bony structure. For example, in the horse-eye jack *Caranx latus*, the distal portion of the cleithrum is cellular, whereas the medial portion is acellular [41]. A detailed comparative analysis of the crystal structure of bone mineral showed that nearly all the mineral crystals from cow, mouse, and chicken bone had dimensions smaller than 35-40 nm, whereas mineral crystals from herring *(Clupea harengus)* acellular bone were as much as three-fold larger [42]. These and ultrastructural data [43] indicate that acellular fish bone accumulates numerous, large crystals well beyond the hole zones, between collagen fibrils. Remodeling types of bone do not appear to accumulate as many crystals beyond the hole zones. Given the importance of bone mineral solubility for calcium homeostasis, the large crystals would be poorly exchanged. This is not a problem for bony fish (teleosts), however, because of a number of important differences between them and mammals [44]. Regulation of plasma calcium content is crucial for all terrestrial vertebrates, and the skeleton is tightly integrated into the animals' calcium homeostasis. Teleosts have a different approach. The absence of osteocytes precludes exchange of mineral by osteocytic osteolysis. Both marine and fresh-water teleosts very efficiently obtain and release calcium from and into the water across their gills and in the intestine.

The skeleton is not needed as a reserve store of calcium [45]. In times of extreme starvation, mineral may be resorbed from the dermal scales rather than bone.

2.1.1.5 Tissue Architecture

Bone tissue can be classified into two microstructural types, depending on its maturity and organization. Newly-formed, immature bone is also known as "woven," and it contains a high ratio of cells-to-matrix and disorganized matrix. Woven bone is characteristic of active periods of growth, fracture callus, heterotopic osteogenesis, and osteosarcoma [46]. With maturation, woven bone is replaced by organized lamellar bone, with a lower cell density and an organized layered pattern of matrix (Figure 2.6).

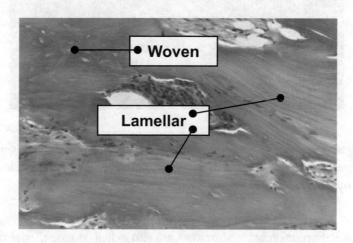

Figure 2.6: Histological section of human cranial bone shows areas with woven and lamellar structures.

Both woven and lamellar bone can be found in the higher level of structural organization of the tissue: trabecular (porous, cancellous, or spongy) and cortical (dense) (Figure 2.7). Trabecular bone is located at the metaphysis and epiphysis of long bones and in cuboid bones (e.g., vertebrae). The porosity of trabecular bone imparts it with a large surface area and a light weight per volume. In histological sections, trabecular bone appears as a network of trabecular struts that are an average 100 to 300 μm thick, with 300-1,500 μm spaces between them [46]. Trabecular bone is characterized by large spatial and temporal variations in porosity, architecture, and mechanical properties. Its mechanical failure is dominated by compressive loads, and it is more compliant than cortical bone. Cortical bone constitutes the diaphysis of long bones, and it can also be found enveloping cuboid bones. Compared with trabecular bone, its density is several folds higher, and its metabolic turnover is several folds lower because of its lower surface area. *In vivo*, cortical bone is subjected to bending, torsional and compressive forces. It is anisotropic, stiff, and, compared with trabecular bone, stronger under compression, but also more brittle. The most complex type of cortical bone is known as

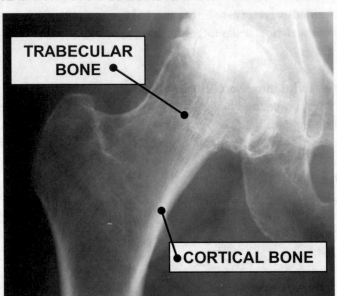

Figure 2.7: Radiograph of proximal femur shows dense cortical bone and the fine structure of trabecular bone. The cortical bone provides resistance to forces and the trabecular bone provides marrow spaces for hematopoiesis and optimal tissue strength for low weight.

Haversian bone. Unit Haversian bone structures are also called "osteons," and they constitute the functional units of mature human bone. Osteons contain a central vascular channel with one or more blood vessels; the channel is surrounded by concentric rings of lamellar bone. The surrounding lamellar bone contains individual lacunae, each of which contains an osteocyte with its extending canaliculi connecting to other osteocytes [6, 46].

Cortical bone may be further characterized as primary or secondary, depending on the mechanism by which it is deposited [47]. Primary bone is formed on pre-existing bone surface. An example of primary bone is that deposited on the periosteal surfaces during growth. Primary bone may be organized as circumferential lamellar bone, or as plexiform bone. Circumferential lamellar bone consists of lamellae are arranged parallel to the bone surface. Blood vessels are located inside the lamellar structure, each surrounded by circular lamellae forming a primary osteon with a Haversian canal at the center (Figure 2.4). Plexiform bone has a different structure and consists of a mixture of layers of woven and lamellar bone. Plexiform bone forms much faster than the more organized circumferential lamellar bone; plexiform structures predominate in fast-growing animals such as cows or racehorses [48].

Secondary bone is that which is generated to replace existing bone in the course of normal remodeling activity. It consists of secondary osteons or Haversian systems that have an approximate diameter of 200 μm and consist of approximately 16 cylindrical lamellae around a central vascular channel. The secondary osteon is separated from the surrounding bone by the cement line (Figure 2.4). Both the cortical and trabecular elements in most adult humans consist of secondary bone [47].

At the organ level, bone is covered by a thin, fibrous, adherent membrane called the periosteum. The periosteum contains two layers: an outer fibrous layer and an inner cambium. The inner cambium is the source of new reactive bone that forms after injury. The inner portion of bone is lined with a fibrous layer called the endosteum, located at the interface of bone with the marrow space. The endosteum also contributes to new bone formation in fracture healing [6, 46].

2.1.1.6 Changes in Structure and Function of Bone with Aging

Human aging is associated with a loss of bone mass throughout the skeleton. Bone loss is due to increased frequency of activation of the bone surface for remodeling, and to an imbalance between bone resorption and formation. Bone loss is partially responsible for the changes in the tissue's architecture that are common in the aging individual. Age-related changes in bone mass, bone mineral density (BMD), and bone architecture depend on various factors such as genetics, gender, anatomic site, and bone type. Evidence indicates that the rates of age-related cortical or trabecular bone loss are faster in women than in men, possibly due to the abrupt loss in estrogen production that accompanies the onset of menopause [49]. Age-related changes in BMD of the lumbar spine (anterior-posterior) for Caucasian women may be described by a peak between the 3rd and 4th decades of life, followed by a 2-4% decrease in the 5th decade, and average decreases of 11%, 18%, and 21% at 55, 65, and 75 years, respectively [50]–[53]. Men, in contrast, maintain peak lumbar spine bone mass through age 50; there may be a 5-6% reduction in the 4th decade of life, and small reductions thereafter [50, 53]. In hips (a common site for age-related fragility fractures) of women, the femoral neck had lost bone mass by approximately 13% at 55 years old, 19% by 65 years old, and 25% by 75 years old [50]–[52], [54, 55]. Studies of other parts of the hip show that Ward's area sustains more dramatic bone loss than the femoral neck, with approximately 12% at age 45 [51]. In men, a small loss of femoral BMD begins to occur between the 3rd and 5th decades of life, and continues without significant acceleration. Studies of femoral bone loss in cohorts from ages 55-90 years old showed that females may lose as much as twice the bone mass lost by men [56, 57]. For measures of the whole skeleton, significant decreases in BMD that average 5%, 10%, and 14% at ages 55, 65, and 75, respectively, have been observed for women [58, 59], whereas for men only small declines of approximately 2% occur every decade from the 4th to the 7th decades of age [60]. Loss of trabecular bone density [61] starts before midlife and continues throughout life [49]. With decreased trabecular bone density, the number and thickness of individual trabeculae decrease, while the size of inter trabecular spaces increases [62, 63]. Cortical bone exhibits age-associated decreases in thickness [64], and increases in porosity [64, 65], which are correlated with decreased BMD [66]. The diameter of the Haversian canals increases with age [64].

Technical advances in imaging methods for bone structure, for example, with quantitative computer tomography (QCT), have allowed for the non-invasive assessment of aging effects on trabecular and cortical bone. It was demonstrated that cortical thickness and BMD in the inferior-anterior portion of the femoral neck in women were relatively preserved between 20 and 90 years of age. A significant age-related decrease in the same parameters was observed in the other three quadrants of the femoral neck [67]. Declines in BMD of the femoral neck, hip, and spine were reported with advancing age, and they were greater in women than in men [68]. In individuals between 67 and 93 years of age, the age-related decrease in bone mass was two- to five-fold greater in women than in men; and accordingly, the age-related loss in bone strength in women was more than two-fold greater than in men [69]. Age-related decreases in bone density and cortical thickness were observed in radii and distal tibias of subjects between 20 and 79 years of age, but trabecular micro-architecture parameters such as number, separation, and thickness of trabeculae were not correlated with age. The changes in cortical bone with age were greater in women than in men after age 50 [70].

The amount and quality of collagen in bone, and the ratio of collagen-to-mineral may also change with aging. Collagen cross-links play a significant role in the biomechanical integrity of bone. Collagen cross-links in bone that are formed enzymatically include lysyl pyridinoline (LP) and hydroxylysyl pyridinoline (HP). LP may have an important function in mineralization because it is almost always confined to calcified tissues (reviewed in [71]). Non-enzymatic glycation-induced cross-links result from the interaction of glucose and collagen [72]. An age-related increase in the non-enzymatic glycation of type I collagen results in accumulation of advanced glycation end products (AGEs) in the organic matrix [73]. Pentosidine (PE) is the most widely investigated non-enzymatic glycation cross-link in bone, present at a concentration of one crosslink per 200-300 collagen molecules [71]. An analysis of the concentration of HP, LP, and PE in 30 human femurs from donors aged 19 to 89 years old showed a significant age-related increase in the concentration of PE, and no age-related changes in the concentrations of HP and LP. AGEs accumulate in the organic matrix of bone, and they are said to inhibit osteoblast proliferation and differentiation, and osteoclast differentiation and function [74, 75]. By modifying the material properties of the organic matrix of bone, AGEs have an effect on bone resistance to fracture. Increased AGEs levels in serum or urine were correlated with increased vertebral fractures in postmenopausal women [76], older diabetic subjects [77] and the general population after adjustment for traditional risk factors [78]. The concentration of PE was inversely correlated with the yield and ultimate strengths and the fracture toughness of the bone [79]. Bisphosphonate therapy for postmenopausal osteoporosis results in the accumulation of AGEs, which, in turn, are correlated with increased bone fragility [80]. Age has reportedly no effect on the concentration of the other cross links, HP and LP, in human bone [80].

A study of the effect of age in human femurs showed a correlation with increased denaturation of collagen in the osteoid, with increased concentration of AGEs, and decreased area of newly formed osteons, but no effects on the number of newly formed osteons [81]. The authors suggested that in aged individuals, the deposition of an abnormal collagen matrix may hamper normal mineralization, leading to reduced quality of newly remodeled bone. In human femoral cortical bone from men between 17 and 73 years of age, primary lamellar bone (which survives turnover for decades) was found to undergo two chronological stages. Until the middle of the third decade of life, the organic matrix became increasingly mineralized with increased mineral crystallinity. Thereafter, neither the mineral content nor crystallinity changed with age. In contrast, in secondary osteons of remodeling, the mineral content did increase with increasing age [82].

Because bone remodeling in older individuals is not as efficient as in younger individuals, there are alterations in the degree of bone mineralization with age. Bone mineralization is a measure of the density of mineral in the matrix and it is related to the material properties of the tissue. Thus, even though many studies show a decrease in BMD (amount of mineral per unit volume) or bone mineral content with age, bone mineralization increases with age [83]. This is possible because the proportion of highly mineralized bone increases [83]. Increased mineralization of bone with age has been attributed to low or incomplete remodeling of the matrix and increased mineral crystallinity [37, 84]. Osteoclasts may be less able to resorb bone with higher density of mineral, as has been shown *in vitro* [85]. Pathologically increased mineralization of the ECM of bone is termed "hypermineralization," and it influences the mechanical properties of the matrix by decreasing the amount of plastic deformation that can occur prior to crack initiation and ultimate failure; the net result is increased brittleness [86, 87]. Accordingly, studies with bones of older osteoporotic women reported the presence of larger mineral crystals and concluded that the hypermineralization impairs the mechanical performance of the bones [88]. Changes in the crystallinity of bone mineral may not be related to size only; some authors report relatively unchanged size of apatite crystals with age [84] and of a change in crystal geometry towards a more spherical one (less elongated) in older osteoporotic individuals; this parameter alters the local stress distributions in the tissue and affects its load bearing properties [37]. Increased porosity in older bone accentuates the effects of hypermineralization because it translates into reduced bone volume and increased fragility [38]. Changes in the nature of the mineralization of bone have been associated with decreased bone fracture toughness in aged individuals [89]. It is likely that a range of mineralization optimizes fracture toughness in human bone [90]. There are some discrepancies in findings about the effect of age on mineralization of bone [91, 92], but they may be due to anatomic location and to whether the bone is cortical or trabecular [93].

Bone cells also undergo age-related changes. In young mice, osteoblasts were arranged in a dense membrane-like structure over all bone surfaces, with a layer of thin collagen microfibers and osteoid [94]. With increasing age, there was reduced collagen accumulation and a progressive reduction in the width of the immature collagen layer and osteoid, until eventual disappearance. The membrane-like arrangement of osteoblasts at the bone surface was lost with age, and this resulted in bone surfaces (as opposed to cells) left vulnerable to physical (loading) or chemical (pH) changes in the environment. Aging osteoblasts are less able to contribute to the maintenance of dense inner lamellar bone in osteons, as was demonstrated in human tibia specimens where the ability of osteoblasts to synthesize matrix proteins and to initiate matrix mineralization became increasingly deficient with age [95]. With age, osteoblasts may lose their ability to respond to two major regulatory factors in skeletal maintenance: vitamin D and TGF-β [96]. The loss in the response of aging osteoblasts to these factors may be related to fewer receptors for vitamin D [97] and decreased expression of TGF-β [98].

Several studies have assessed age-related osteocyte death, which is usually quantified as the percentage of empty lacunae in histological specimens. Some report progressively increased osteocyte death with age, with one of the pioneer reports by Frost [99]. An age-related reduction from 100% osteocyte viability at 25 years to 74% in the eighth decade of life was reported for the femoral head; this decrease was correlated with the number of adjacent microfractures [100]. Subsequently, a similar effect was reported for femoral cortical bone, with the additional insight that the coefficient of variation of the number of osteocyte lacunae also increased with age [101]. In contrast, osteocyte viability was not correlated with age in bone from the human mandible [102] or from iliac crest [103]. Age had no effect on the density of osteocyte lacunae in femoral head trabecular bone until 70 years, at which point there was a sudden sharp loss of osteocytes [104]. It is likely that the correlation between osteocyte viability and age may depend on the anatomical location of the measurement [105]. Age-related effects on the number and viability of osteocytes in lacunae also appear to depend on the type of bone (cortical or trabecular). Some investigators conclude that the presence of osteocytes renders bone more easily resorbable, and therefore, as the individual ages, selective resorption may result in decreased density of osteocyte lacunae [106].

In vitro studies reveal that bone-resorbing osteoclasts also undergo age-related changes. Osteoblast- and stromal cell-induced osteoclastogenesis, and the number of osteoclast precursors increase with age of mice [107], and with precursor cells from humans [108].

2.1.2   FUNCTION

Bone serves two major, yet distinct roles: one is mechanical and the other is metabolic. Mechanically, bone provides support and locomotion to the organism while affording protection to vital internal organs. In order for bone to fulfill these mechanical functions, it needs strength and elasticity. The metabolic function of bone concerns its contribution to calcium homeostasis because the mineral phase can be mobilized when necessary.

2.1.2.1   Bone Remodeling

Bone undergoes continuous and extensive cell-mediated remodeling in the adult skeleton. Portions of bone are resorbed by osteoclasts and subsequently replaced with new bone secreted by osteoblasts in a process of cell and matrix renewal. Cortical bone remodeling takes place at the apex of tunnels, or cutting cones, and the cycles of resorption and formation are highly orchestrated so that the mechanical integrity of the tissue is not compromised during its renewal.

As a result of the cellular activities in bone remodeling, it is possible to repair day-to-day microdamage and to renew minerals so that they are more soluble and accessible for ion homeostasis [46]. It is this mechanism of bone turnover that confers bone with the ability to heal fractures.

Continuous bone remodeling also enables bone to incorporate and integrate metallic implants, bone grafts, and other osteocompatible materials (Figure 2.8).

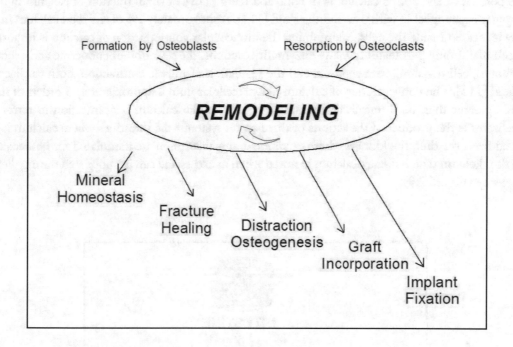

Figure 2.8: Adult human bone is characterized by a remodeling process in which bone mineral and matrix are resorbed focally by osteoclasts and replaced by osteoblasts. This continuous cellular activity contributes to mineral homeostasis and allows for fracture healing and incorporation of bone grafts and implants.

Bone formation and resorption can be monitored with biomarkers in serum or urine. Markers of bone formation include bone-specific serum alkaline phosphatase, osteocalcin, and collagen type I propeptides. Markers of bone resorption include urinary pyridinium cross-links, cross-linked peptides and hydroxyproline and serum tartrate-resistant acid phosphatase [46].

After age 25-30 years, there is a slight shift in the balance between bone resorption and formation, such that the resorption of bone outpaces its formation, which may result in progressive bone loss [109]. The decline in bone's ability to remodel efficiently has been evidenced by increased amounts of microdamage in bone with increasing age [110]. The rate of increased microdamage is significantly higher in females than in males [110]. Menopause is associated with an increase in bone remodeling activities and contributes to risk of osteoporosis [64, 111].

2.1.2.2 Calcium Homeostasis

Of the inorganic elements in the human body, calcium is the fifth most abundant [112]. Ninety-nine percent of the body's calcium is in bone and teeth [113]. A small fraction of calcium in bone is rapidly exchangeable with extracellular fluid [114]. Approximately 1% of the total calcium in the body is located inside the cells. Maintaining the intracellular concentration of calcium is important for cell membrane permeability, many enzymatic reactions, the secretion of endocrine and exocrine hormones, cell division, cell apoptosis, and the regulation of muscle contraction, both cardiac and skeletal [112]. The concentration of calcium in extracellular fluid is approximately 5 orders of magnitude greater than its intracellular counterpart. Extracellular calcium concentration in terrestrial vertebrates is the product of the actions of three organ systems: the intestines where calcium is absorbed from the diet, the kidneys where excretion and reabsorption are modulated for homeostasis, and the skeleton where dietary calcium is stored when in excess and can be mobilized during dietary deficiency (Figure 2.9).

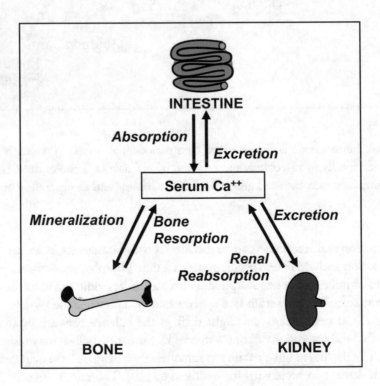

Figure 2.9: Three organ systems contribute to maintenance of the concentration of calcium in extracellular fluids. Dietary calcium is absorbed in the intestine, stored in the skeleton, and excess is excreted in feces and urine. During dietary deficiency, calcium is mobilized from bone and reabsorbed in the kidney.

A small percent of ingested calcium is absorbed in adults and the percentage decreases with age [115]. As much as 60% of ingested calcium may be absorbed in infants and young children, compared with 15%-20% in normal adults. Normal individuals on an average diet may ingest 1 gram calcium/day, but absorb only about 20% in the duodenum and upper jejunum; the rest is excreted in the stool, at a loss of 700-800 mg per day. The age-related decrease in the rate of ingested calcium absorption in adults is approximately 0.2% per year; additionally, there is a decrease of 2% in the perimenopausal period [116]. In a study that measured intestinal calcium absorption in 262 healthy postmenopausal women ages 40-87 years, absorption in women older than 75 years was merely 28% of that in women younger than 75 [117]. Similarly, a 5% decrease in calcium absorption was reported in women with low calcium intake who were 85 years of age and older, compared with those who were 69-84 years old [118].

Calcium homeostasis in the extracellular fluid is tightly controlled by the synchronized actions of parathyroid hormone (PTH), calcitonin, and vitamin D_3 metabolites. Working in concert, these hormones regulate the concentration of calcium in serum by acting on bone, intestine, and kidney. Dietary calcium enters the blood by intestinal absorption and it is excreted via the kidneys and/or stored in bone. PTH is synthesized by the parathyroid gland, and is released when a decrease in serum calcium signals its calcium-sensing receptor [119]. PTH acts to maintain serum calcium ion concentrations within a narrow physiological range. In bone, PTH stimulates acute efflux of calcium and more gradually stimulates bone resorption. In the kidneys, PTH stimulates calcium reabsorption. By stimulating the activation of vitamin D, PTH indirectly acts on the intestine to increase calcium absorption [120].

Calcitonin is the anti-hypercalcemic hormone that increases in blood after a calcium-rich meal. It inhibits renal reabsorption of calcium and osteoclastic bone resorption, thereby decreasing risk of post-prandial hypercalcemia [121].

Vitamin D is a calcitropic hormone. In its active form, it enhances the intestinal and renal absorption of calcium as well as its resorption from bone. There are two major sources of vitamin D_3. It is produced in the skin by conversion of 7-dehydrocholesterol into cholecalciferol (vitamin D_3) upon the action of ultraviolet light, hence its name the sunshine vitamin. It is also available from diet and vitamin supplements. Vitamin D_2 (ergocalciferol) is a common food supplement. Both D_3 and D_2 are hydroxylated to 25-hydroxyvitamin D (25(OH)vitamin $D_{2,3}$) in the liver, and they are relatively inactive precursors until, in response to PTH or to low calcium or phosphorus levels, they become hydroxylated to active 1,25-dihydroxyvitamin $D_{2,3}$ in the kidneys [122] (Figure 2.10).

The ability of the kidney and intestine to maintain proper calcium homeostasis declines with age, with the net result of a loss of stored calcium from the bone [46]. The decrease in the ability of the kidneys and intestine to maintain calcium homeostasis is related to changes or deficiencies in the vitamin D/PTH regulatory system. Vitamin D deficiency is more prevalent than previously expected in the United States. Although many dairy foods are supplemented with vitamin D, intake and skin synthesis are not sufficient. Serum vitamin D levels in the order of <10-15 ng/mL are generally considered as inadequate for overall health in healthy adults [123]. Holick [124] recently

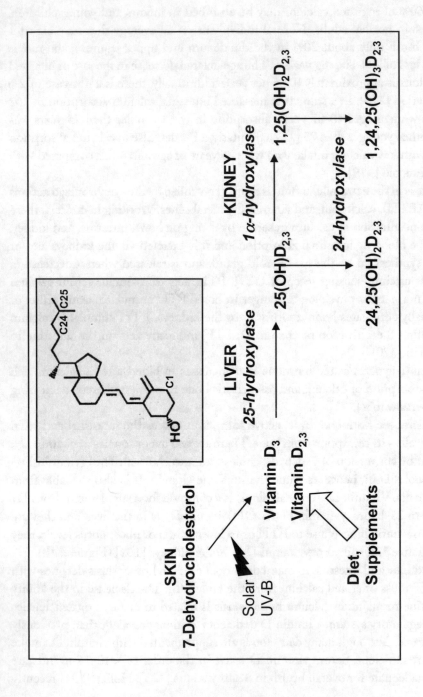

Figure 2.10: Metabolism of vitamin D. Inset shows the structure of vitamin D_3 and the numbered carbons that are hydroxylated by tissue hydroxylases. Solar UV-B light (290–315 nm) converts 7-dehydrocholesterol in skin to vitamin D_3, which is released into the blood. Endogenous and ingested D_3 and its analogue D_2 are transported to the liver by serum albumin and binding proteins for hydroxylation of the carbon-25. This is followed by 1 α-hydroxylation in the kidney to the active form $1,25(OH)_2 D_{2,3}$. These reactions convert the fat-soluble precursor into more water-soluble forms with increased specificity for the vitamin D receptor in target tissues. High levels of $1,25(OH)_2 D_{2,3}$ are associated with risk of hypercalcemia and are prevented by 24-hydroxylation.

indicated the magnitude of the problem. Vitamin D deficiency or insufficiency affects approximately 1 billion people worldwide, 40 to 100% of elderly American and European individuals not living in nursing homes, more than 50% of postmenopausal women taking osteoporosis medication, 52% of Hispanic and black adolescents in Boston, 48% of white preadolescent girls in Maine, 42% of black women between 15 and 49 years old, 32% healthy students, physicians, and residents at a Boston hospital, 30 to 50% of children and adults in sunny countries where most of the skin is shielded from the sun such as Saudi Arabia, United Arab Emirates, Australia, Turkey, India and Lebanon. Even in women who took extra measures to increase vitamin D uptake during pregnancy, 73% of them and 80% of their infants were vitamin D deficient at the time of the birth. The evidence for prevalent vitamin D insufficiency portrays an overwhelming world-wide public health problem. Vitamin D deficiency is due to multiple factors: an intake that is lower than the needed levels, limited exposure to sunlight, an inability of the kidneys to convert vitamin D to its active form, or inadequate absorption of vitamin D from the digestive tract. There is new understanding about the amounts of vitamin D needed for vitamin D sufficiency. The Institutes of Medicine estimated that adequate daily intake of vitamin D for individuals between 1 and 70 years of age is 600 international units (IU) [125]. The requirements for people older than 70 years is 800 IU per day [125] because there is a decrease in the dietary absorption of vitamin D and its conversion to the active metabolites [126]. In addition, there is an age-related increase in intestinal resistance to the calcitropic actions of the active form of vitamin D [127]. Recent practice has evolved to recommend a daily intake of 1000-2000 IU for adults.

In contrast, serum PTH levels increase with advanced age [128], with a prevalence of 20-60% [129]. The age-related increase in serum PTH is likely a secondary effect of vitamin D deficiency and resultant decreased levels of calcium and phosphate in serum [124]. Although increased PTH stimulates renal activation of vitamin D, it also stimulates osteoclastogenesis that can lead to bone loss and even hypercalcemia in the extreme [130].

2.1.3   MECHANICAL PROPERTIES

The fundamental concepts of mechanics of bone and the various mechanical tests used to assess them are thoroughly explained by Wang and colleagues in their lecture on Fundamental Biomechanics in Bone Tissue Engineering that is part of this series [131]. Most mechanical testing of bone involves specimens of standard shape subjected to controlled mechanical and environmental conditions. Laboratory tests are designed to eliminate the effects of the bone geometry. One of the widely used categories of tests for the mechanical properties of bone involves the measurement of the relationship between load and deformation. Load-deformation tests typically entail uniaxial loading of the bone specimen under tension or compression, or less often torsion loading. The tests may be controlled for load or for strain (deformation). In a load-control test, predetermined loads are applied and the specimen's deformations are measured and recorded. In a strain-control test, the specimen is compressed or elongated to a specific percentage of its length (strain), and the equilibrium load that results is recorded. Data are presented as a stress vs. strain plot (Figure 2.11 (top)), where stress is the normal force (in tension or compression) divided by the cross-sectional area of the specimen, and strain is the deformation (elongation or shortening) divided by the original length of the specimen. The initial region of the stress-strain curve is linear, and its slope is called the Young's modulus, E, which gives an index of a material's stiffness [47]. The Young's modulus of cortical bone is approximately 17 GPa, and although there is a large variation for trabecular bone, a relatively stiff sample may exhibit a Young's modulus on the order of 1 GPa [132]. This can be compared to the Young's modulus for 316L stainless steel (E=200 GPa), a material commonly used with bone plates, as a reference [132]. Beyond the linear region of the stress-strain curve, the yield region represents the beginning of permanent deformation of the specimen. The yield strength is the stress at which microstructural damage leading to permanent deformation begins to occur. The ultimate strength, also simply referred to as the "strength" of the bone specimen, is the stress at failure (fracture) [47]. Whereas uniaxial loading tests evaluate stress and strain to determine the Young's modulus and the strength of bone, torsion tests record the values of torsion and angular twist, which when normalized to the appropriate geometric parameters yield a shear stress-strain curve whose slope is the shear modulus, G. More complex tests involve multi-axial loading configurations, with loads acting simultaneously in various different directions, or cyclic loading at various amplitudes. Cyclic tests run at low amplitudes of loading allow for the determination of fatigue parameters, which are important for structural integrity upon *in vivo* day-to-day loading conditions.

The mechanical properties of bone are defined by the direction of the applied loads, by the anatomic location, and by the type of bone [46]. The parallel orientation of collagen fibers with respect to the long bone axis makes the bone stronger upon compression along its axis (longitudinal) than transverse to its axis. This phenomenon is called anisotropy [47]. Because cortical and trabecular bone have different micro-structures, in particular when it relates to porosity and density, they have different material properties. The apparent density of bone is the ratio between the mass of the bone specimen and its bulk volume (including bone marrow spaces). The typical ranges of apparent densities for trabecular and cortical bone are 1.0-1.4 and 1.8-2.0 g/cm^3, respectively [47]. The density of trabecular bone has a much larger standard deviation, and, therefore, its mechanical properties also exhibit a larger standard deviation when compared with cortical bone. Bone mineral density has been considered a predictor of bone strength [133]. Most studies indicate, however, that the inclusion of microarchitectural parameters can further increase the correlation between the mineral density and the strength of trabecular or cortical bone [134, 135], although there are some contradictory data [136]. Loss of connectivity and removal of trabeculae result in a significant decrease in trabecular bone strength, particularly in the case of osteoporosis [137]. Micro-structural parameters of trabecular bone determined by magnetic resonance imaging, such as bone volume fraction (bone volume per trabecular volume), spacing, number and thickness of trabeculae significantly correlate to Young's modulus and compressive strength [138, 139]. The porosity of cortical bone plays a major role in its mechanical properties. For example, cortical bone porosity is correlated with Poisson's ratio in rat femurs [140], its square root is inversely correlated with the elastic modulus of human femoral necks [141], and a second degree polynomial regression model correlates it with elastic modulus

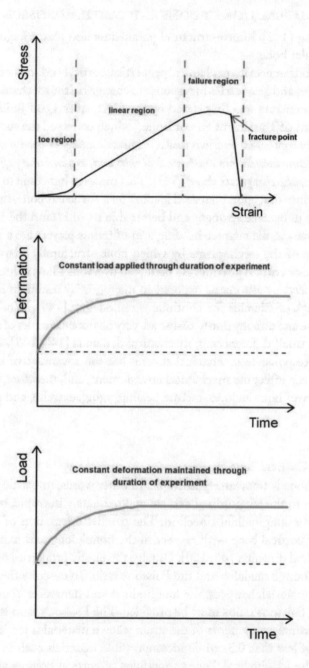

Figure 2.11: (top) Schematic of a stress-strain curve with its toe, linear, and failure regions. (middle) In a creep experiment, a constant load is applied to the tissue. Its deformation increases until it reaches a stable value. (bottom) The stress-relaxation experiment is the inverse: a constant deformation is applied and the stress increases until it reaches a stable value.

upon transverse loading [142]. Micro-structural parameters also play a major role in determining the strength of trabecular bone.

In spite of their differences in mechanical properties, cortical and trabecular bone are similar in adapting their properties and geometries in response to changes in the mechanical demands imparted upon them. This phenomenon was first stated by the German surgeon Julius Wolff in 1892, in a work entitled: "The Law of Transformation of Bone," which conveyed the succinct observation that "*as a consequence of primary shape variations and continuous loading, or even due to loading alone, bone changes its inner architecture according to mathematical rules and, as a secondary effect and governed by the same mathematical rules, also changes its shape*"[143]. This message has come to be known as "Wolff's law." Wolff's law describes that upon increased loading on a particular bone, the bone remodels itself in shape and structure to become stronger and better able to withstand the high stresses imparted upon it. The dense bones in the racquet-holding arm of tennis players are a good example of such adaptation [144]. One of the mechanisms by which bone structurally adapts to high loads is by thickening of the exterior cortical portion. As predicted by Wolff's law, reduced loading conditions such as those experienced by astronauts exposed to microgravity results in loss of bone mass and strength due to the lack of stimulus for continuous remodeling [145]. While hypertrophy due to weight-bearing exercise and atrophy due to disuse are very obvious examples of adaptation [146, 147], bone also responds to small differences in mechanical demands [148, 149]. The ability of bone to dynamically adapt in response to mechanical stresses has the advantage of maintaining structural integrity. Factors that can affect the mechanical environment, and, therefore, induce changes in the structure and geometry of bone include fracture healing, aging, exercise, and surgery.

2.1.3.1 Properties of Cortical Bone

Elasticity.   Cortical bone is transversely isotropic. In other words, its elastic properties for loading in the plane transverse to the longitudinal axis are approximately isotropic, but very different from those for loading in the longitudinal direction. The parallel orientation of collagen fibers in the mineralized matrix of cortical bone with respect to the bone's long axis makes bone stronger and stiffer in the longitudinal direction [47, 150]. Usually, the elastic properties of an isotropic material are described by the Young's modulus and the Poisson's ratio. To describe the elastic properties of a transversely isotropic material, however, the longitudinal and transverse Young's moduli, the shear modulus, and the two Poisson's ratios must be provided. The Poisson's ratio is the ratio between the transverse and longitudinal components of the strain when a material is loaded. Ordinary materials have Poisson's ratios of less than 0.5, while incompressible materials such as liquids have Poisson's ratios of 0.5 [47]. The longitudinal Young's modulus of cortical bone is approximately 1.5-fold greater than its transverse Young's modulus, and 5-fold its shear modulus. The Poisson ratio of human cortical bone has been estimated as 0.39 [151]. The strength anisotropy of human femoral Haversian bone (the difference between longitudinal and transverse strength) is more pronounced upon tension than upon compression (Table 2.1).

Table 2.1: Mechanical properties of human femoral Haversian bone under tension and compression loading in the longitudinal or transverse directions. Adapted from [47]; data from ([a][152]; [b][151]).

	Tension		Compression	
	Longitudinal	Transverse	Longitudinal	Transverse
E (GPa)[a]	17.9	10.1	18.2	11.7
Strength (MPa)[b]	133	51	195	133

Factors that affect the elastic properties of cortical bone may be intrinsic, such as porosity and mineralization (which together determine apparent density), architecture, organization of collagen fibers (lamellar bone with its parallel collagen fiber arrangement is stronger than woven disorganized bone), or they may be extrinsic such as fatigue damage (which is detrimental to mechanical properties) and rate of deformation [47].

Strength.   The strength of cortical bone depends on the anatomical source, the type of loading (tension, compression, or torsion), and also, due to its anisotropy, on the direction of loading. Human cortical bone is stronger with compression than with tension for both longitudinal and transverse loading (Table 2.1). It has been suggested that cortical bone adapts to daily demanding activities whereas compressive loading predominates over tensile loading, for example, during walking.

Viscoelastic behavior.   Cortical bone behaves as a viscoelastic material. Viscoelastic materials exhibit both viscous and elastic characteristics when undergoing deformation. The resulting response to stress is a time-dependent strain that slowly reaches equilibrium. This is also known as "creep." During typical moderate daily activities such as slow walking and slow running, the strain rate of cortical bone can vary by more than one order of magnitude, resulting in variations of 15% in the Young's modulus. The range of strain-stress rates for cortical bone, which maximizes energy absorption, is 0.01 to 0.1 per second. The yield and ultimate strengths of cortical bone increase with the strain rate, which can be interpreted as cortical bone becoming stronger during strenuous activity. At extremely high strain rates in the order of >0.1 per second, which characterize high impact trauma, cortical bone is more brittle [132].

A plot of strain vs. time for a constant stress level is called a creep curve (Figure 2.11 (middle)). The initial portion of a creep curve shows a rapid increase in strain. Eventually, the rate of increase in strain diminishes until a constant creep rate is reached. Prolonged application of the stress results in a marked increase in the creep rate until there is fracture. Creep fracture occurs at certain levels of loading for cortical bone, and it occurs faster for higher stress levels. Cortical bone has a higher resistance to compressive creep fracture than to tensile creep fracture [132].

Fatigue.   The fatigue properties of bone are evaluated by placing a specimen under cyclic loading at specified stress levels, and recording the number of cycles before the specimen fractures. The observations are usually presented graphically with the stress level on the y axis, and the logarithm of the number of cycles before fracture (also known as "fatigue life") on the x axis. The fatigue damage mechanism involves crack initiation, crack propagation, and final fracture. Although continuous bone remodeling interrupts the growth of cracks as they are repaired, small cracks often unite into bigger cracks. Bigger cracks may result in fracture. The mechanisms of formation and propagation of fatigue cracks, and of eventual failure in cortical bone depend on the type of loading. Uniaxial loading is more likely to generate fatigue damage when the stress is applied in tension than when applied in compression. With tensile loading, cracks initiate easily but have difficulty propagating; bone eventually fractures due to micro crack coalescence. On the other hand, with compressive loading, cracks do not initiate easily but their initiation is followed by sudden fracture. In other words, cracks propagate easily [153].

The mechanical properties of human cortical bone progressively deteriorate with aging [154]. Changes include a reduction in material stiffness and strength and an increase in brittleness. The decline in the mechanical properties of cortical bone that occurs with age is attributed to bone loss, increased porosity, hypermineralization [86, 91], changes in the collagen network, and accumulation of fatigue damage (reviewed in [38, 104]).

The age-related decrease in the longitudinal Young's modulus and tensile yield strength of human femoral diaphyseal bone is estimated as 2% per decade after age 20; the age-related reduction in energy absorption is approximately 7% per decade [132]. Ultimate tensile stress, ultimate tensile strain, and energy absorption decrease by 5, 9, and 12% per decade, respectively [93]. The decline in cortical bone strength after age 35 is in the range of 2-5% per decade of age (reviewed in [38]). The age-related deterioration in the elastic properties of cortical bone facilitate the creation of macrocracks and their propagation through the tissue [89].

The fragility of aging cortical bone may also be related to changes in the bone collagen's chemistry [155]. Aging has been correlated with the deposition of an abnormal collagen matrix [81]. The concentration of reducible collagen cross-links in cortical bone was shown to decrease steeply between birth and 25 years, but to persist thereafter [156]. There is some evidence that reduced concentrations of collagen cross-links in cortical bone are associated with reduced bone strength [80].

2.1.3.2 Properties of Trabecular Bone

The mechanical properties of trabecular bone can be studied with stereologic descriptions of its architecture and multivariate regressions of both architectural and density parameters with material properties [157, 158]. Because of the large spatial and temporal variations in the density of trabecular bone, its material properties can vary by more than an order of magnitude (Table 2.2).

Further, the material properties of trabecular bone depend on its architecture, with a very wide range of porosities.

Elasticity.   Trabecular bone is anisotropic, with the orientation of individual trabeculae affecting its elastic modulus by as much as one order of magnitude. The Young's modulus can range from 10 to 2,000 MPa, depending on age and anatomic site. Trabecular bone is significantly less stiff than

Table 2.2: The compressive strength and stiffness of human trabecular bone [47].

Ultimate stress (MPa)	Modulus (MPa)	Sample information	Reference
1.86-1.37	90-70	Lumbar vertebrae of individuals aged 40-60 years	[159]
2.54 ± 0.62	272 ± 195	Lumbar vertebrae cylinders of men aged 54-90 years	[160]
5.60 ± 3.80	424 ± 208	Distal femur cubes	[161]
2.22 ± 1.42	489 ± 331	Proximal tibia of individuals aged 42-76 years	[162]
10.16 ± 8.92	597.9 ± 401.6	Proximal femur, proximal tibia, distal radii and lumber spine of individuals aged 30 to 89 years	[163]
3.21 ± 2.13	376 ± 347	Distal femur of individuals aged 45 to 92 years	[164]

cortical bone [132]. In some cases, trabecular bone contracts upon compression (negative Poisson's ratio). The modulus and the compressive strength in any loading direction may be related to apparent density (mineralized bone mass/bulk volume) and ash apparent density (ash mass/bulk volume) by a power law relationship of the form $y = ax^b$, where y is either the elastic modulus or the compressive strength, x is the parameter apparent density of ash apparent density, and a and b are experimentally-derived constants [165]. The correlation between the compressive strength and elastic modulus of trabecular bone with apparent density and ash density was demonstrated in animal models or human samples [166, 167]. The accuracy of the power law model that correlates apparent density/ash apparent density with the elastic modulus/compressive strength of trabecular bone may further benefit from the inclusion of the parameters of bone volume fraction (bone volume/bulk volume), and ash fraction (ash mass/dry bone mass), because both parameters independently influence both apparent density and ash apparent density [168].

A typical compressive stress-strain curve for trabecular bone begins like one for cortical bone, with a region of linear slope (Figure 2.12). With further compression, the specimen reaches its yield and ultimate stress points, similar to cortical bone. Unlike cortical bone, however, the load does not fall to zero past the ultimate stress because trabecular bone does not fracture into pieces. Rather the trabeculae compact against each other, and the load is maintained or can rise [47]. The linear portion of the stress-strain curve for trabecular bone behaves similar for tension and compression, but the post-yield behaviors are completely different. Failure occurs by fracture of individual trabeculae; this reduces the tissue's ability to support compressive loads until complete fracture occurs. Trabecular bone loaded with tension past the ultimate strength cannot support any further load; all the load transfers to the surrounding tissue, causing further fractures through crack propagation [132].

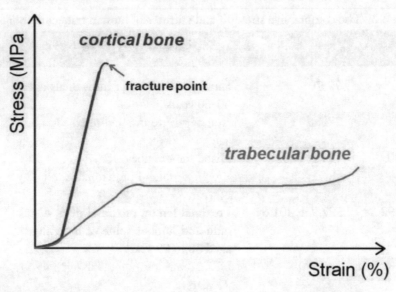

Figure 2.12: Comparison of typical compressive stress-strain curves for trabecular and cortical bone. Both curves start with a region of linear slope. After further compression, cortical bone fractures and the load goes to zero. In contrast, trabecular bone does not fracture into pieces but rather the trabeculae compact against each other, and the load is maintained or can rise [47].

Although some studies indicate that the stress at failure for cancellous bone is the same with tension as compression [169, 170], others find that trabecular bone is stronger with compression [171]. These discrepancies may be related to the anatomical source of the tested specimens, such that trabecular bone in some locations of the body may be adapted to withstand compressive loads.

Viscoelastic behavior. Although the viscoelastic behavior of human trabecular bone is poorly understood, there is evidence from bovine and, to a lesser degree, human specimens that suggests that trabecular bone exhibits significant time-dependent behavior in the form of creep or relaxation [172]–[174]. The compressive strength, stiffness, and ultimate strain of human trabecular bone from the proximal tibia have shown linear and power-law correlation to strain rates between 0.0001 and $10 \, s^{-1}$ [175]. A pioneer study with bovine trabecular bone specimens showed that the creep behavior of trabecular and cortical specimens were qualitatively similar, and that power law relationships existed between the stress normalized by cross-sectional area and the creep rate at steady state [172].

Fatigue. The majority of human bone fatigue analyses available in the literature concern cortical bone [176]. An early study showed that trabecular bone was less resistant to fatigue damage than cortical bone [177]. The fatigue behavior of trabecular bone may be similar for different anatomical sites and species, and the dominant failure mechanisms may occur at the ultrastructural level [178]. There is a relationship between microarchitecture and the fatigue behavior of trabecular bone with compression [179]. The fatigue life of human trabecular bone from lumbar and thoracic vertebrae with axial compressive loading is well predicted by the applied stress normalized by the specimen's volume fraction [179]. Fatigue may lead to fracture of individual trabeculae. Such microcracks can be repaired by micro-callus formation if the dimensions are small. Fatigue damage to trabecular bone inflicted by low cycle loading is due to fracture and buckling of individual trabeculae [176], whereas that inflicted by high cycle loading is due to crack accumulation/propagation [180]. There are limited experimental data on human trabecular bone fatigue. Simulation studies of trabecular bone fatigue damage confirm that the complete fracture of trabecular struts is not an absolute requirement for the failure of high-density trabecular bone, and that when cracks develop in a relatively low number of trabecular struts, there is a subsequent relatively large reduction in the specimen modulus that can rapidly lead to fracture [180, 181].

Effects of aging.   Fractures of trabecular bone occur more frequently in aged individuals and in post-menopausal women in particular. The most prevalent fracture sites are the spine, distal radius, and proximal femur (hip) [182]–[186]. An estimated 16% of Caucasian women will fracture a distal radius at some point after 50 years of age [187]. This is in contrast to only 2-3% for Caucasian men. The incidence of wrist fractures increased by 6-fold with age for women over 45 years old and remained unchanged for men. The incidence of hip fractures increased exponentially with age for both sexes, but was always 2-fold lower in men compared with women [188]. An investigation of the association of a history of Colles distal radius or spine fractures with remaining lifetime risks for hip fracture found differences with age and gender. A history of spine fracture in postmenopausal women was associated with a higher risk (15%) of sustaining a hip fracture in the remaining lifetime than was a history of Colles fracture (13%) or of no fractures (9%). In contrast, men over 50 years of age with a history of Colles fracture, spine fracture, or no fracture had 8%, 6% and 3% risks, respectively, of future hip fracture [189].

The age-related changes in the density and architecture of trabecular bone, in particular age-related bone loss, lead to a deterioration of its mechanical properties [86, 92, 190]. For example, the compressive strength of human trabecular femoral bone decreased by 8.5% per decade of age in subjects aged 20 to 102 years of age. In those specimens, surface-to-volume ratio and mean separation of the trabecular plates increased with age, while mean thickness and connectivity of the trabecular plates decreased. The structural age-related changes had a profound impact in the apparent density of the bone samples, which in turn accounted for 92% of the reduction in mechanical strength [190]. In samples from human proximal tibial trabecular bone from donors between 16 to 83 years, Young's modulus and ultimate stress were highest between 40 and 50 years of age, whereas ultimate strain and failure energy showed maxima at younger ages. The age-related reductions in Young's modulus, and ultimate stress were non-linear, and were believed to be mainly a consequence of trabecular bone loss, rather than a decrease in the quality of trabecular bone [92].

Changes in the metabolism of bone collagen in relation to the age-related loss of trabecular bone mass and strength are often overlooked. A study of iliac crest bone biopsies from healthy women and men aged between 18 and 96 years showed an age-related decline in the intrinsic collagen content but no biochemical changes in the bone collagen with age. It was suggested that decreased bone collagen content with age was indicative of increased mineralization. The changes in collagen and mineral content were correlated with declined mechanical properties of the bone specimens [86].

2.2   ARTICULAR CARTILAGE

2.2.1   STRUCTURE

There are three types of cartilage in the body: hyaline articular cartilage, elastic cartilage, and fibrocartilage. Hyaline cartilage is also called articular cartilage and is found on the articulating surfaces of bones, costal cartilages, larger cartilages of the larynx and nose, the cartilages of the trachea and bronchi and the growth plate. Its ECM is composed mainly of type II collagen and proteoglycans. Hyaline cartilage is shiny and has a pearly blue color. Elastic cartilage is rich in the protein elastin and is present in the outer ear, larynx, and epiglottis. It contains many bundles of yellowish elastic fibers lying in a solid matrix. Elastic cartilage is more elastic and resilient than hyaline cartilage or fibrocartilage. Fibrocartilage is a mixture of fibrous and cartilaginous tissues in various proportions. It contains mainly type I collagen and proteoglycans, with dry weight percentages for these components, respectively, higher and lower than those of hyaline cartilage. Fibrocartilage is found in the annulus fibrosus of intervetebral discs, the menisci of the knees, the symphysis, the marginal plates and interarticular discs of some joints such as the TMJ, the sesamoids of some cartilages, and the lining of bony grooves for tendons. It occurs in bulky masses. It is tougher and less resilient than articular cartilage, but its function as a loading bearing material is not satisfactory [191, 192]. This monograph focuses on articular cartilage.

2.2.1.1   Tissue Composition

Articular cartilage tissue is 65-80% fluid, 20-35% ECM comprised mainly of collagen and proteoglycans, and a very small population of sparse, but somewhat uniformly distributed and highly specialized cells known as chondrocytes. The shape of the chondrocytes, the biochemical composition of the ECM, and the spatial distribution and organization of the components of articular cartilage vary throughout the depth of the tissue, measured from the articulating surface to the cartilage-bone interface. Based on these differences, cartilage can be divided into four zones: superficial, transitional (or middle), deep, and calcified (Figure 2.13) [3].

2.2.1.2   Articular Cartilage Zones

The region of articular cartilage closest to the joint space is known as the superficial zone. It contains the highest concentration of collagen and the lowest concentration of proteoglycans. The collagen fibrils in the superficial zone are thin and arranged parallel to the articulating surface. The chondrocytes are elongated, and flattened in shape (10×2-3 μm), with their long axis oriented parallel

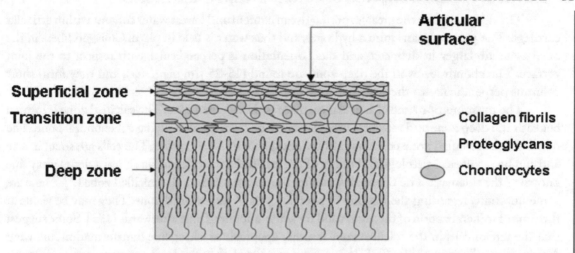

Figure 2.13: Schematic of the zones of articular cartilage. In the superficial zone, chondrocytes and collagen fibers are oriented tangential to the articulating surface. This zone exhibits the highest tensile strength. In the transition zone, cell and collagen fiber alignment are random. In the deep zone, chondrocytes and collagen fibers are oriented perpendicular to the articulating surface. The space inside the collagen network is filled with a negatively charged proteoglycan-water phase that maintains the collagen network in constant tension.

to the articulating surface [3, 193], and they are stiffer than chondrocytes from the middle and deep zones [194]. The superficial zone acts as a barrier from the synovial fluid by maintaining tissue hydration and preventing the entrance of unwanted materials and the exit of vital cartilage components. The small size estimated for pores in the superficial zone (50Å) suggests that they should allow for the transport of only very low molecular weight solutes from the synovial fluid (<20 kD) into the tissue. Small molecules such as urea, amino acids, and ions are able to penetrate the cartilage matrix from the synovial fluid. Large molecules such as albumin should be virtually excluded from freely diffusing into the tissue, and their diffusion rates should be further retarded due to friction with the matrix proteoglycan macromolecules. It has been suggested that cartilage derives some of its nutrition from the synovial fluid through a "physiological pump" when cartilage is compressed under load and fluid is exuded from the matrix. Subsequent relaxation leads to reabsorption of the fluid, which has been enriched with synovial fluid nutrients. It has been shown, however, that the "physiological pump" has a significant effect in increasing the transport of large solutes such as serum albumin, but not on small solutes such as glucose [3, 193].

The next zone is the transitional zone. Collagen fibers in the transitional zone of cartilage are less organized and have a larger diameter than those in the superficial zone. The chondrocytes in the transitional zone are round with an approximate diameter of 15-25 μm, and they do not follow any apparent organization [3, 193].

The deep zone has the greatest proteoglycan content and lowest water content within articular cartilage. The proteoglycans form a hydrogel, and thus water is held in place. Collagen fibers in the deep zone are larger in diameter, and their orientation is perpendicular with respect to the joint surface. The chondrocytes in the deep zone are round (15-25 μm diameter), and they form short columns perpendicular to the articulating surface [3].

The most interior zone of articular cartilage is the calcified zone. The calcified zone is located beneath the deep zone and separates articular cartilage from the underlying subchondral bone. The matrix in the calcified zone contains calcium phosphate mineral crystals. The cells are small in size and can be pyknotic. Histological staining of the calcified zone of cartilage shows a dense wavy line known as the tidemark. The tidemark separates the deep zone from the calcified zone [3]. There are some questions regarding the viability of chondrocytes in the calcified zone. They may be viable at the time of mineralization of the matrix and for some amount of time afterwards [195]. Some suggest that the fate of cells in the calcified zone is either apoptosis or osteogenic transformation following matrix mineralization [196, 197]. It has been suggested that the local environment at different micro-anatomical sites may determine whether some hypertrophic chondrocytes transform into bone cells [198].

The differences in chondrocyte shapes in the different articular cartilage zones have been demonstrated to relate to differences in metabolism and differentiation [199]. These differences are often maintained during *in vitro* culture. Cultured cells from the upper (superficial) zone deposited less proteoglycans than those from the lower zones, and those from the calcified zone were distinguished by expression of alkaline phosphatase [200]. Chondrocytes from the superficial zone produced less cartilage-specific extracellular matrix in three-dimensional *in vitro* culture when compared with chondrocytes from the middle zone [201]. Self-assembled high density cultures of chondrocytes from the middle and deep zones resulted in constructs with distinct regions and 250% more GAGs per dry weight and more collagen per dry weight than cultures of superficial zone chondrocytes [202].

2.2.1.3 Zones of Extracellular Matrix

Another classification for the ECM of articular cartilage concerns its proximity to the chondrocytes. According to this classification, three ECM regions exist: pericellular, territorial, and interterritorial [3]. These regions vary in composition and in the diameter and organization of their collagen fibers. Pericellular matrix is found adjacent to the chondrocyte membrane and envelopes the whole chondrocyte. The pericellular matrix is thin, is rich in proteoglycans, hyaluronan, and noncollagenous proteins [193], and contains organized type IV collagen [203].

The territorial ECM surrounds the pericellular matrix and contains thin fibrils of collagen type II that form a distinct basket-like network enveloping the individual cells [3, 193]. The term "chondron" was first used by Benninghoff in the 1920s to refer to the complex formed by a chondrocyte and its pericellular and territorial matrices. It was suggested that a chondron is the primary functional and metabolic unit of cartilage matrix homeostasis [204]. Chondrons are thought to protect the cells by dampening mechanical, physicochemical, and osmotic changes during dynamic loading [5, 205].

The interterritorial ECM is the largest of all regions, and it provides the majority of articular cartilage's material properties. It is the zone between chondrons. The interterritorial matrix contains the largest collagen fibers and the majority of proteoglycans [3].

2.2.1.4 Chondrocytes

Chondrocytes are highly differentiated and specialized cells responsible for the normal synthesis and maintenance of articular cartilage. They originate from mesenchymal condensations during development. They secrete large amounts of ECM during growth, and they maintain the ECM of mature articular cartilage. The cellular density of cartilage decreases with age. In humans, chondrocyte density has been estimated as $2\text{-}3 \times 10^6$ cells/mm^3 in young embryonic cartilage, 10^5 cells/mm^3 in newborn femoral condylar articular cartilage, and one tenth of that in young adults [193].

Because cartilage lacks a blood supply, chondrocytes exist in a relatively low oxygen environment (approximately 6% in the surface zone, and 1% in the deep zone of human femoral head cartilage), use the anaerobic pathway for energy production, and derive most of their nutrition and waste exchange from mass transfer with the synovial fluid (reviewed by [193]). Chondrocytes do not make intercellular contacts, and rarely engage in mitotic activity after the cessation of skeletal growth. Nonetheless, chondrocytes are permanently active in cartilage turnover by replacing short-lived proteoglycans, and producing collagen, core protein, and degradative enzymes such as collagenase, gelatinase, stromelysin, and cathepsins. The maintenance of a healthy articular cartilage ECM depends on the ability of the chondrocytes to coordinate and balance the synthesis of matrix components, their incorporation into the ECM, and the degradation and release of ECM components. Chondrocytes maintain this delicate balance *in vivo* and *in vitro* in response to cues in the chemical and mechanical environments in the form of soluble mediators (Table 2.3), compressive loading [206, 207], hydrostatic pressure [208], and/or electric fields [209]. Chondrocytes respond to those forms of mechanical stimuli by altering their synthetic activity in a manner that depends on the type, timing, magnitude, and mode of the load.

Chondrocytes in osteoarthritic cartilage behave differently from normal adult chondrocytes. They differ in cell number, metabolic activity, and cellular phenotype. Changes in the number of chondrocytes in osteoarthritic cartilage may occur either through increased cell death or cell proliferation. Increased cell death in osteoarthritic cartilage is a topic of controversy, with some reports refuting increased apoptosis [229] and some suggesting it [230]. There is reproducible histopathological evidence of cell proliferation in clusters that may be a localized attempt at repairing and regenerating the damaged tissue [231]. The ability of osteoarthritic chondrocytes to maintain a balance between anabolic (matrix producing) and catabolic (matrix degrading) activities is compromised in osteoarthritis. In fact, early in the onset of the disease, there is increased synthetic activity [232]. Nevertheless, chondrocytes fail to compensate for the damage in the matrix. Further, chondrocytes have a direct role in the degradation of the matrix by up-regulating the expression of

Table 2.3: Some soluble mediators that affect articular cartilage homeostasis.

Soluble mediator	Effect on articular cartilage cells and matrix	Representative references
Transforming growth factor β (TGF-β)	Elicits seemingly opposing effects depending on the experimental conditions. Examples: inhibits IL-1 induces degradation of articular cartilage, stimulates synthesis and prevents loss of proteoglycans, stimulates DNA synthesis, inhibits proteoglycan deposition, mitigates or potentiates the matrix-degrading effects of retinoic acid.	[210]–[215]
Insulin-like growth factor I (IGF-I)	Mitigates degradative effects of cytokines. Stimulates proteoglycan and collagen type II deposition. Stimulates mitotic activity.	[212], [216]–[218]
Fibroblast growth factor (FGF)	Mitogen. Potentiates both anabolic and catabolic functions. Potentiates the effects of IL1. Mitigates the effects of IGF-I and TGF-β	[218]–[221]
Interleukin-1α (IL-1α)	Induces increased degradation and decreased synthesis of proteoglycans; decreases synthesis and degradation of type II collagen. Suppresses TGF-β signaling.	[212, 213], [217], [222, 223]
Tumor necrosis factor α (TNF-α)	Induces increased degradation and decreased synthesis of proteoglycans. Suppresses TGF-β signaling.	[212, 224]
Retinol	Induces matrix degradation, inhibits proteoglycan synthesis and/or sulphate incorporation.	[225, 226]
Retinoic acid	Induces matrix degradation. Depresses proteoglycan anabolism.	[215], [227, 228]

degradative enzymes such as matrix metalloproteinases (reviewed in [233]). In a study with human articular cartilage that was discarded during knee arthroplasty and separated into areas with different degrees of degeneration, aggrecan gene expression was found to be inversely correlated with disease severity [234]. Other changes in chondrocyte phenotype as a result of osteoarthritis include the presence of collagens not normally found in adult articular cartilage such as type X collagen and other chondrocyte differentiation genes [235].

2.2.1.5 Fluid

Up to 80% of healthy articular cartilage is aqueous, most of it located in the molecular pore space of the ECM, with a one-third occupying the collagen interfibrillar space, and a small fraction located in the intracellular space. The fluid content of articular cartilage decreases from the surface through the deep zone. The interstitial fluid in articular cartilage is mostly water, with some dissolved inorganic salts [3], and the electrolytes sodium, calcium, potassium, and chloride [236]. It has low viscosity, (1.2 relative viscosity), and low protein and sulfur contents (indicating low content of ECM components) [236]. The electrolytes in the interstitial fluid interact with the negatively charged sulfated proteoglycans; this interaction generates a solid proteoglycan-water gel entrapped within the collagenous network. The interaction between proteoglycans and interstitial fluid generates internal pressurization through Donan osmotic pressure. The internal pressure swells the tissue and contributes to resistance against compressive loads [237]. Interstitial fluid can flow from the tissue upon application of a pressure gradient across the tissue, or deformation of the solid matrix phase. The fluid flows out of the tissue, as has been visualized in the form of exudation and collection of droplets of fluid on the surface of cartilage specimens loaded in tension or compression [47]. The movement of interstitial fluid through the molecular size pores of the ECM generates high frictional resistance, thus resulting in low permeability. The flow of fluid through the tissue helps supply nutrients, substrates, waste, and soluble factors that aid cartilage matrix homeostasis [193].

2.2.1.6 Collagen

Collagen comprises approximately 50% of the dry weight of articular cartilage. Collagen provides articular cartilage with tensile strength and immobilizes the proteoglycans within the ECM. Collagen in the ECM of articular cartilage exists in the form of relatively thin fibers that are cross-linked into a 3D network. The diameter of the collagen fibers ranges from 10 to 100 nm, and it may increase with age and disease [3]. The collagen fibers in articular cartilage are organized into an arcade-like arrangement, such that they follow the direction tangential to the articular surface in the superficial zone, are randomly organized in the transition zone, and are perpendicular to the articular surface in the deep zone.

Approximately 90% to 95% of articular cartilage's collagen is type II, and the rest is types V, VI, IX, X, and XI [3, 238]. Collagen type II is a homotrimeric molecule composed of 3 identical α chains. The highly conserved antigenic epitopes in the type II collagen molecule confer it with unique antigenic properties when compared with other fibrillar collagen types: antibodies against chicken type II antibodies will react with type II collagens from all other species, including human, calf, mouse, rat, sheep, dog and shark [238]. Collagen types IV and IX function to maintain the association between type II collagen and proteoglycans in the ECM [46]. Types IX and XI comprise 5-10% of the collagen in the heterotypic collagen fibers of articular cartilage's ECM [239].

Enzymatic breakdown of collagen is catalyzed by matrix metalloproteinases. The cartilage-specific type II collagen is considerably less soluble and less easily degraded than type I of bone [46]. Although proteolysis of collagen in articular cartilage is normally slow, it can be accelerated during repair, remodeling, and/or degeneration. The half life of collagen has been calculated as more than 200 years in humans [5].

2.2.1.7 Proteoglycans

Proteoglycans consist of a protein core with attached polysaccharide chains that are called glycosaminoglycans (GAGs). Approximately 90% of the proteoglycans in articular cartilage exist as large complexes known as aggrecans. Aggrecans contain a large protein core (approximately 250 kDa). Up to 100 chondroitin sulfate chains (approximately 20 kDa each), and 50-60 keratan sulfate GAG chains (approximately 5-15 kDa) attach to the protein core, as do some small oligosaccharides (Figure 2.14).

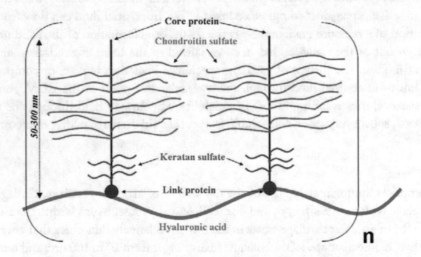

Figure 2.14: Figure 14 Schematic of a proteoglycan aggregate. A protein core (black) binds through link protein (red) to a large hyaluronic acid chain (green). The hyaluronic acid-core protein bond mediated by the link protein is almost irreversible at physiological conditions of temperature and pH. Chondroitin sulfate (orange) and keratan sulfate (blue) chains attach to the protein core, as do some small oligosaccharides (not shown). The protein core-glycosaminoglycanoligosaccharide complex is known as an aggrecan. A hyaluronic acid chain may bind as many as 200 aggrecans.

Hyaluronan binds to one of the ends of the protein core. The hyaluronate-core protein bond is mediated by link protein, and it is almost irreversible at physiological temperature and pH. A single hyaluronan chain can bind as many as 200 aggrecans, resulting in a large proteoglycan aggregate. These very large aggregates associate with water and become immobilized within the collagen network of articular cartilage [3, 240].

Articular cartilage also contains smaller proteoglycans. These include biglycan and decorin, which are dermatan sulfate proteoglycans. These provide a small but significant contribution to the mass of proteoglycans in articular cartilage. They are associated with collagen fibrils, and they bind fibronectin, growth factors, and other macromolecules. Although it is established that biglycan and decorin fulfill different functions in the ECM, their individual functions are not well understood. Their binding to collagen fibrils inhibits fibrillogenesis, whereas their adhesion to fibronectin inhibits cell adhesion, and their binding to TGF-β inhibits its mitogenic activity [240].

GAGs are long-chain, unbranched repeating disaccharide units with carboxyl and/or sulfate groups along their chains. These groups become ionized in solution (COO^- and SO^{3-}). In the physiologic environment, these anions are countered by Ca^{2+}, Na^+, and other ions. In order to balance the negative charges in the proteoglycans, a large number of ions must be mobilized from the exterior fluid and into the tissue, resulting in a large difference in ion concentrations between the inside and the outside of the tissue. The gradient in ion concentrations has the effect of drawing water into the tissue, which results in swelling and expansion, and places the collagen network under tension. The tension in the collagen network balances the swelling pressure, thus providing the tissue with excellent compressive resilience [240].

Articular cartilage contains three major types of GAGs: chondroitin 4- and 6-sulfates, keratan sulfate, and dermatan sulfate. The 4- and 6-chondroitin sulfates are the most prevalent, accounting for 55% to 90% of the total GAGs in articular cartilage, depending on factors such as age, disease, and anatomical location [3]. Although it is clear that the synthesis of GAG chains is affected by many factors, the relationship between the GAG composition of aggrecan, and the age and anatomical site of the tissue is not well understood [240].

Proteoglycans are produced intracellularly in the endoplasmic reticulum. The protein core is synthesized on ribosomes and transported to the Golgi complex, where the GAG chains are added and where the proteoglycan is enzymatically elongated. Proteoglycan molecules are then immediately secreted into the ECM, in other words, aggrecan is not stored inside the cells. Link protein is synthesized by the same intracellular pathway; whereas hyaluronan is formed by synthase enzymes at the plasma membrane, so that secretion into the ECM immediately follows synthesis. Aggregation of the proteoglycans occurs outside of the cell. Many factors including age, disease, and anatomical location alter the quality, rate, and amount of proteoglycans secreted by the chondrocytes. Proteoglycans are continuously degraded enzymatically and released from articular cartilage as a part of normal tissue maintenance. The half-life of large aggrecan monomers has been estimated as 3.4 years, whereas the half-life of the hyaluronan-binding domain may be on the order of 19-25 years [241, 242]. In degenerative conditions, such as osteoarthritis, the degradation of proteoglycans occurs at an accelerated rate. The proteolytic enzymes that degrade proteoglycans, namely metalloproteinases and aggrecanases, are elevated in the inflammatory responses that accompany joint disorders such as rheumatoid arthritis; this elevation results in increased proteoglycan degradation [243].

2.2.1.8 Changes in Structure and Function of Articular Cartilage with Aging

Changes in articular cartilage with aging include a reduction in its thickness [244], decreased cell density, lower mitotic and overall activity of the chondrocytes [245], and changes in the concentration, size, and distribution of proteoglycans [246, 247]. Both the protein core and the GAG chains diminish in length, and proteoglycan aggregation decreases with advancing age [246]. All of these changes affect the mechanical properties of the tissue. Although there are reports of age-related losses of mechanical properties in articular cartilage [248], these studies are influenced by the high prevalence of osteoarthritis with increasing age, and it is unclear whether these changes in mechanical properties are specific to normal aging or a consequence of the disease.

2.2.2   FUNCTION

2.2.2.1   Mechanical

Articular chondrocytes are highly sensitive to changes in their mechanical environment, such as changes in pressure or deformation of the matrix. These changes occur with loading or unloading of the joints. Upon sensing these mechanical changes in their microenvironment, articular chondrocytes initiate biochemical modification of the ECM by synthesizing or degrading its macromolecules [249]. The term "mechanotransduction" refers to the mechanism(s) by which cells convert mechanical information into cellular activity. It concerns the mechanisms by which cells detect changes in their mechanical environment and respond by modifying their gene expression patterns and their synthesis and breakdown of ECM. In articular chondrocytes, mechanotransduction is a new field of study and not entirely well understood, although it is clear that it plays a crucial role in the development of articular cartilage regeneration therapies. Recent research efforts suggest that articular chondrocytes' response to mechanical stimuli is mediated by signaling pathways involving $\alpha5\beta1$ integrin as a mechanoreceptor [250], and results in direct changes in transcription, translation, and post-translational modifications [251], as well as in changes in cell-mediated synthesis/degradation of matrix [252]. These biochemical modifications in turn affect the tissue's mechanical properties.

Articular cartilage is subjected to high loads during normal and/or strenuous activity. The average and peak pressures on the hip joint during walking, descending and climbing stairs have been estimated in the order of 4.4 to 5.0 MPa and 10.8-12.7 MPa, respectively [253]; more strenuous activity such as running at 3.5 meter per second results in a contact force of 5.2 ± 0.4 times body weight on the hip joint during the push-off phase [254]. Loads are applied statically (such as when sitting or standing) or cyclically (such as when walking, jumping, or running). The application of loads occurs repetitively and continuously for many decades. Articular cartilage is therefore constantly stressed and under a high demand for mechanical performance.

Another function of articular cartilage is that of providing lubrication for joint articulation [255]. Articular cartilage performs this function so efficiently that the friction coefficients of synovial joints have been compared to those of an ice skate on ice [47]. Upon application of a load onto articular cartilage, water flows from the cartilage in response to the interstitial pressurization. The extruded water associates with lubricant molecules, resulting in a lubricative complex. This complex prevents direct cartilage-to-cartilage contact during joint movement. Upon withdrawal of the load, the pressure decreases, and water is re-absorbed to the proteoglycans in the matrix [46, 255]. This method of lubrication is believed to be the main mechanism of lubrication for synovial joints [47, 255]. Boundary lubrication in articular cartilage is mediated by proteoglycan 4 (PRG4), also known as lubricin, superficial cartilage zone protein, or megakaryocyte stimulating factor precursor [256], and is enhanced by hyaluronan [257]. Lubricin is found in the synovial fluid and in the superficial zone of articular cartilage [258]. Although some of the protein accumulates in the superficial zone of cartilage [258], most is secreted into the synovial fluid by chondrocytes of the superficial zone, and also by synoviocytes and meniscal cells [259].

2.2.2.2 Metabolism

Immature cartilage obtains its nutrition from both the synovial fluid and the subchondral bone, but in adult cartilage, due to the tidemark and the calcified zone, most of its nutrients are derived from the synovial fluid [3, 260]. The synovial fluid coats the articular surfaces of the joint with a very thin layer (for example, 10-20 mm in the knee joint). Theoretically, given the small size of the pores of the superficial zone of articular cartilage, only low molecular weight (20 kD) components of the synovial fluid should be able to diffuse into the cartilage. In fact, however, and dependent on the integrity of the tissue, large molecules such as interleukin-1, prostaglandins, some growth factors, and proteoglycan fragments can move in and out of the tissue [3, 261]. These observations suggests that a small fraction of relatively large pores exist in cartilage through which the transport of large molecules takes place [261]. It was also shown that the deep and calcified layers are permeable to small solutes; therefore, the subchondral circulation may contribute to the nutrition of deep cartilage [262]. Static compression of articular cartilage decreases the diffusivity of solutes (3-40 kDa) through the tissue [263], whereas dynamic loading increases solute transport (0.5-10 kDa) by enhancing convection [264].

Chondrocytes both secrete ECM and mediate its enzymatic degradation. Because many of the matrix components of articular cartilage are tissue-specific, there is interest in measuring their degradation products in plasma, urine, or synovial fluid as biomarkers of articular cartilage turnover [265, 266].

2.2.3   MECHANICAL PROPERTIES

The mechanical behavior of articular cartilage is best described with a biphasic model, where the two phases are a fluid and a solid, and they are both incompressible [236, 267]. The fluid phase comprises the water and ions that constitute 65-80% of the tissue, whereas the solid phase is the ECM structure, made mainly of collagen and proteoglycans. The theory assumes that cartilage ECM is a soft, porous, and elastic solid, and that its pores communicate and are filled with water [47]. A key feature is that when a load is applied to cartilage, the solid and fluid phases respond differently. The solid phase undergoes deformation, which generates a pressure gradient through the tissue. The fluid phase flows through the pores as a result of the pressure gradient, and some of it dissociates from the hydrogel complex and is temporarily exuded. The biochemical composition and the ultrastructural organization of the solid phase play vital roles in articular cartilage's ability to sustain the high loads that it experiences daily. The mechanical properties of articular cartilage vary even at different points within the same anatomical structure. Some average literature values for the human knee joint are presented in Table 2.4.

2.2.3.1 Permeability

The solid and fluid phases of articular cartilage share the loads applied onto the tissue. Partitioning of the load between the solid and fluid phases at the articular surface shields the solid matrix from excessive stress and improves lubrication [271]. Upon application of a pressure gradient, water flows through the pores of articular cartilage ECM. This flow of water is a result of the compaction of the matrix, which raises the pressure in the interstitium of the collagen fibers. Normal articular cartilage has a permeability coefficient ranging from 10^{-15} to 10^{-16} m^4/Ns. This low permeability indicates that very high pressure gradients are necessary to move water through articular cartilage [3]. Therefore, most of the total applied load is supported by the fluid phase through pressurization of the entrapped fluid. Over 75% of loads applied to articular cartilage are supported by the fluid phase [272, 273]. Characteristics of the loading regime such as type and rate of loading have an effect on the proportion of the load supported by the fluid phase. For example, the contribution of the fluid phase to load support increases at higher loading frequencies during dynamic compression [274], and at higher applied loads during frictionless rolling contact [273]. In sum, permeability plays an important role in the load support function of articular cartilage.

The permeability of articular cartilage changes with the deformation that occurs in the tissue upon application of a load [275]. Permeability decreases with compression because as the fluid flows out of the tissue there is a reduction in the tissue's water content and porosity, both of which have a net effect of increasing the density of negative charges in the interstitium. The deformation-dependent decrease in the permeability of articular cartilage has been modeled as exponential [276]. The strain-dependence of permeability effectively prevents the tissue fluid to be rapidly and excessively exuded fluid exudation upon application of a compressive load, promotes interstitial fluid pressurization for load support, and dissipates energy during cyclic loading [3].

2.2.3.2 Compression

Articular cartilage behaves as a viscoelastic material, in other words, its behavior varies with time upon application of a constant load or deformation. Two types of experiments are used to define the properties of a viscoelastic material (Figure 2.11 (middle), (bottom)). A creep experiment consists on applying a constant compressive load that gradually deforms articular cartilage deforms such that the strain increases with time until equilibrium (Figure 2.11 (middle)). In a stress-relaxation experiment, articular cartilage is compressed to a constant strain; the stress increases rapidly to a peak value, after which, it gradually decreases to an equilibrium value [47] (Figure 2.11 (bottom)). The compressive modulus of the solid matrix of normal articular cartilage ranges from approximately 0.4 to 1.5 MPa, depending on anatomic location [3].

Table 2.4: Average values of the Poisson's ratio, compressive modulus and permeability coefficient for the cartilage of the human femoral condyle (FC) and patellar groove (PG). n/a = not available.

Poisson's ratio		Compressive modulus (MPa)		Permeability coefficient ($m^4 N^{-1} s^{-1}$)		Reference
FC	PG	FC	PG	FC	PG	
0.098 ±0.069	0	0.701 ±0.228	0.530 ±0.094	(1.182 ±0.207) ×10^{-15}	(2.173 ±0.730) ×10^{-15}	[268]
n/a	0.158 ±0.148	n/a	0.845 ±0.383	n/a	1.8 ×10^{-15}	[269]
0	0	0.60	0.42	1.48 ×10^{-15}	2.46 ×10^{-15}	[270]

2.2.3.3 Shear

Collagen fibrils in the transitional and deep zones of articular cartilage are in a random spatial arrangement. The fibers are stretched when a shear load is applied to the articular surface, and the entrapped proteoglycan molecules are sheared. Shear loads do not result in interstitial fluid flow because there are no pressure gradients or volume changes. A dynamic sinusoidal shear experiment measures the magnitude of the dynamic shear modulus and tan δ. The dynamic shear modulus may be interpreted as the intrinsic shear stiffness of the solid matrix at any given loading frequency; it ranges from 0.2 to 2.5 MPa in the frequency range 0.01 to 20 Hz in normal human articular cartilage. Tan δ represents the viscoelastic energy dissipation in articular cartilage during a cycle of shear; δ of $0°$ means no energy dissipation (the material is purely elastic), whereas δ of $90°$ means the material is purely viscous material and highly dissipative; δg or articular cartilage ranges from $20°$ to $9°$ with increasing frequency. Articular cartilage resists shear better when it is compressed [3]. The shear modulus varies by up to two orders of magnitude within the same anatomical location, depending on the depth of the tissue [277].

2.2.3.4 Tension

Articular cartilage responds to tensile loads by flow-dependent and flow-independent viscoelastic mechanisms. With tension, it is the solid phase of cartilage that carries most of the load, in particular, the collagen fibers. The typical tensile stress-strain curve for articular cartilage starts with a toe region with a low modulus followed by a linear increase in the modulus. The slope of the linear portion is termed the tensile modulus or tensile stiffness. In the linear portion of the curve, the collagen fibers are stretched straight (Figure 2.15).

The tensile modulus reflects the stiffness of the collagen network when in tension, and it ranges between 5 to 50 MPa, depending on various factors such as location, depth, orientation, and tissue composition. The superficial zone has the highest tensile modulus due to its higher concentration and degree of orientation of collagen. The tensile modulus is correlated with collagen content and with the ratio of collagen to proteoglycan content. Normal human articular cartilage is capable of withstanding tensile stresses, ranging from 9 to 18 MPa, and tensile strains from 60% to 120% [278]–[280].

2.2.3.5 Swelling

The proteoglycans in articular cartilage are rich in negatively charged groups that are spaced 10 to 15Å apart and that confer a field charge density that ranges from 0.1 to 0.15 mEq/ml at normal physiologic conditions. Each negative charge is countered by a cation from the external fluid environment. The mechanism for maintaining electrical neutrality results in a total concentration of ions inside the tissue that is much larger than that in the external fluid environment. This difference in ion concentrations generates a swelling pressure inside the tissue that is greater than the pressure in the external bath [3]. For example, normal articular cartilage immersed in saline swells by 2-4% or less [281], and under physiologic conditions its swelling pressure is 0.1-0.25 MPa [3]. Articu-

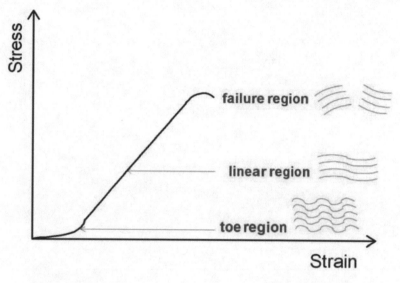

Figure 2.15: Schematic of a tensile stress vs. strain curve for collagenous tissue such as articular cartilage. The collagen fibers become straight in the toe region, are stretched in the direction of the load in the linear region, and rupture in the failure region (Adapted from [47]).

lar cartilage swelling generates stresses in the solid matrix. The swelling pressure is the sum of an osmotic pressure p (generated by the association of the fluid's counter ions with the tissue's fixed charge density) and a chemical-expansion stress that results from large charge-to-charge repulsive forces. The solid matrix expands and imbibes water in response to swelling pressure. Under normal conditions, the elastic stiffness of the matrix can resist the swelling pressure. When there are changes in the collagen or proteoglycan components of the matrix, however, changes in swelling behavior appear. Osteoarthritic cartilage, for example, can swell by as much as 10% [282]. Damage to the collagen network results in increased swelling [283], with a linear relationship reported between the percentage of degraded type II collagen molecules and swelling [284], whereas loss of proteoglycans results in decreased swelling pressure [285]. Excessive swelling of articular cartilage results in diminished mechanical properties, increased permeability, and an increased tendency to injury [285, 286]. Swelling pressure significantly contributes to load support with light compressive loads, but for high loads or dynamic compressive loads, interstitial fluid pressurization is the major contributor to load support.

CHAPTER 3

Development

3.1 INTRAMEMBRANOUS BONE FORMATION

In intramembranous bone formation, there is differentiation of mesenchymal tissue directly into bone. In the initial stage of intramembranous bone formation, mesenchymal tissue condenses, and mesenchymal cells differentiate into osteoblasts that produce collagenous extracellular matrix with irregularly oriented fibers. The osteoblasts condense and many become surrounded with secreted osteoid. They further differentiate into osteocytes in lacunae that interconnect with other cells. The ostcoid mineralizes, and the first bony islands or spicules appear. The spicules fuse together, forming immature trabeculae which are to be replaced during secondary bone formation by mature trabeculae. Intramembranous bone formation is the mechanism of formation of some of the bones of the skull (such as the mandible and the skull vault), the flat bones of the face, most of the clavicle bones, the terminal phalanges and the periosteal bony band surrounding the shaft of long bones [287].

3.2 ENDOCHONDRAL BONE FORMATION

The term endochondral means "from within cartilage." In endochondral bone formation, mesenchymal cells differentiate into chondrocytes, which mature through well orchestrated phases of cellular proliferation, hypertrophic differentiation, apoptosis, vascularization, and replacement of cartilage with bone. A "growth plate" emerges, on which osteoblasts differentiate and initiate the process of ossification. Endochondral bone formation occurs in embryogenesis and in fracture callus. Growth and differentiation factors that play an important role in the regulation of cell proliferation, differentiation, and apoptosis during endochondral bone formation include bone morphogenetic proteins (BMPs), fibroblast growth factors (FGF), parathyroid hormone-related peptide, and members of the Hedgehog and Wnt families. These factors signal the developing mesenchymal cells and chondrocytes, inducing changes in gene expression that are mediated by transcription factors such as Smads, Msx2, Sox9, signal transducer and activator of transcription (STAT), and core-binding factor alpha 1 (cbfa-1) (reviewed by [288]).

Information is available about the time course of ossification in humans. Bones form from a cluster of mesenchymal cells during the development of embryo and fetus. At the 6th week of embryonic life, mesenchymal cells condense and differentiate into chondrocytes. The cells at the periphery of the condensation form a perichondrium. The cells synthesize a cartilaginous model for the future skeleton, roughly in the same shape of the future skeletal elements, called cartilaginous anlage [289]. After the cartilage anlage is formed, chondrocytes at its center undergo terminal differentiation into hypertrophic chondrocytes, aided by Cbfa-1 [290]. Hypertrophic chondrocytes direct the mineralization of the ECM surrounding them, secrete vascular endothelial growth factor (VEGF) and other factors, attract matrix-digesting cells (chondroclasts), and direct adjacent perichondrial cells towards osteoblastic transformation. Hypertrophic chondrocytes then undergo apoptosis. The secreted VEGF promotes the invasion of the tissue by blood vessels [291]. The matrix left behind by the hypertrophic chondrocytes provides a scaffold for the invading osteoblasts to deposit a true bone matrix within it, also called primary spongiosa. During the 7th week of embryonic life, the transformed perichondrial cells deposit a layer of bone via intramembranous bone formation; this layer of bone envelops the periphery of the cartilage anlage. At the end of the 8th week of embryonic life, capillary buds invade the central portion of the anlage, signaling the transition from embryonic into fetal development. The invading blood vessels bring mesenchymal cells into the anlage that differentiate into osteoblasts and osteoclasts (for reviews, see: [288, 292]). The osteoblasts produce an osteoid matrix of primary trabeculae on the surfaces of the calcified cartilaginous bars. Osteoclasts remove these primary trabeculae and a medullary canal expands. This combination of endochondral bone formation by osteoblasts and osteoclastic resorption results in the formation of a growth plate, at either end of the bone. The growth plate, also called physis, is a cartilaginous zone located between the shaft and the epiphysis or bone end. For reference, the metaphysis is the part of the bone immediately beneath the growth plate, and the diaphysis is the mid shaft region between two growth plates. The chondrocytes in the growth plate are organized in vertical columns, which grow appositionally, thereby constituting the functional units of longitudinal bone growth [293]. New bone formation occurs on the metaphyseal side of each growth plate (Figure 3.1) [294]. At about the time of birth, secondary centers of ossification forms at the ends (epiphysis) of many bones [295]. The secondary ossification center comprises an epiphyseal growth plate that continues to elongate and enlarge the bone. Bone elongation stops when the epiphyseal plates ossify [296]. The epiphyseal end of a long bone enlarges due to radial apposition of cells around the secondary ossification center, and the bone grows around the secondary ossification center much slower than around the primary growth plate. The secondary center of ossification determines the shape and form of the joint surface for diarthrodial joints (reviewed in [297]).

The growth plate allows for elongation and widening of a bone with minimal risk of fracture. Growth plate cartilage is unlike articular cartilage in its blood supply, which confers it with unique biochemical properties, and its architecture. The growth plate becomes replaced by bone tissue until it is completely obliterated and elongation ceases. The synthesis and degradation of cartilage matrix in the growth plate remain balanced to ensure that the height of the growth plate remains approximately constant [298]. The balance in synthesis and degradation of matrix is regulated by endogenous enzymes such as collagenase, gelatinase, and stromelysin. These enzymes are activated by interleukin-1 and plasmin and their activity is regulated by tissue inhibitor of metalloproteinases [299].

The cartilaginous component of the growth plate exhibits three distinct zones: reserve, proliferative, and hypertrophic. Longitudinal growth of the bone is possible because cells proliferate on the epiphyseal face of the growth plate, and the cartilage becomes mineralized on the metaphyseal face of the growth plate [47].

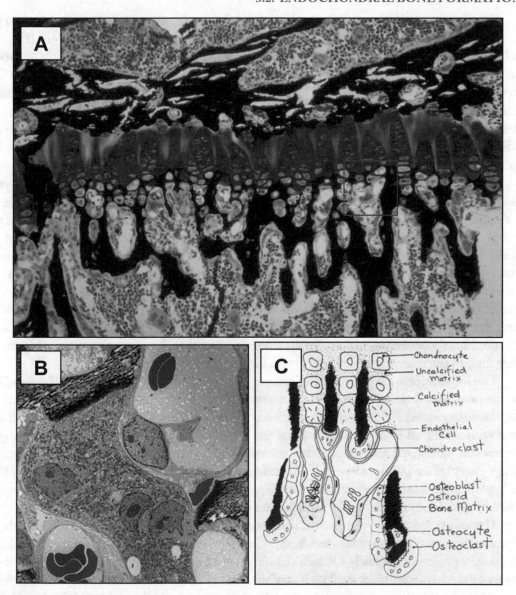

Figure 3.1: Anatomy of Growth Plate. (a) Photomicrograph of histological section of rat proximal femur shows purple zone with columns of chondrocytes, mineralized matrix (black), and metaphysis with multinucleated chondro/osteoclasts and blood-filled (pink erythrocytes) capillaries (von Kossa/H&E stain). (b) Electron micrograph shows the ultrastructure of multinucleated chondro/osteoclast with ruffled border opposing the cartilage matrix and adjacent capillaries and erythrocytes. (c) The ultrastructural organization of an area like that in the red box is illustrated.

The reserve cartilage zone has also been termed as stem, terminal, or resting zone, and it is located below the secondary center of ossification, most distal to the diaphisis of the bone (Hansson LI). Daily growth in length of diaphysis can be monitored with oxytetracycline [300]. The growth zone is narrow, and its contours are irregular. It contains sparse cells that divide rarely [300]. The reserve zone becomes obliterated with growth of bone from the secondary centers of ossification towards the periphery (reviewed by [294]). Compared with other zones of the growth plate, the matrix of the reserve zone has the highest collagen type II content. The collagen fibers in the matrix are disorganized, and there are aggregated proteoglycans, which inhibit matrix mineralization. The oxygen tension is low (20.5±2.1 mm Hg) [297].

The proliferative cartilage zone is responsible for longitudinal growth. It contains elongated cells that proliferate rapidly and are arranged in orderly longitudinal columns [301]. The uppermost cell in each column is the progenitor cell for longitudinal growth of the column. The total number of progenitor cell divisions is determined genetically for each growth plate. The generation of one new chondrocyte in the proliferation zone results in a net average gain in longitudinal bone growth of the new chondrocyte's height, approximately 7 μm; each proliferating chondrocyte produces a net matrix volume equivalent to its own volume, or approximately 2000 μm^3 [301, 302]. These cells have a high capacity for protein synthesis and exhibit the highest rate of proteoglycan metabolism within the growth plate. The matrix is similar to that in the reserve zone in that collagen fibrils are not uniformly organized, and proteoglycans are aggregated. The oxygen tension in this zone is higher than in the reserve zone (57.0±5.8 mm Hg) [297].

The hypertrophic cartilage zone of the growth plate is the site of matrix calcification. This zone contains cells that are enlarged to approximately 5 times the size of those in the proliferative zone; the enlargement is due to cytoplasmic and nuclear swelling. These cells are surrounded by a matrix with decreased interterritorial volume and increased collagen content [303]. Hypertrophic chondrocytes participate actively in matrix mineralization, as evidenced by their high synthesis of alkaline phosphatase [304], proteases [305] and type X collagen [306], and type II collagen [307]. The hypertrophic zone is avascular, has low oxygen tension (29.3±2.4 mm Hg), and the highest content of glycolytic enzymes in the growth plate [297].

The mineralization of the growth plate is different from that which occurs in adult bone. The mitochondria of hypertrophic chondrocytes in the growth plate have high intracellular calcium content that decreases gradually towards the direction of matrix calcification in the lower chondrocytes of the hypertrophic zone. The loss of mitochondrial calcium in the lower chondrocytes is accompanied by the appearance of matrix vesicles that accumulate calcium [308] (Figure 3.2). The formation of the mineral crystals starts at these matrix vesicles, which are round or oval structures with an approximate diameter of 100-250 nm [287]. Matrix vesicles are extruded as buds from the plasma membranes of the hypertrophic chondrocytes and act as nuclei for matrix calcification both during development [309], and in the osteoid of post-fetal bone [310]. They contain alkaline phosphatase and organic phosphates in the vesicle membrane that initiate mineral deposition, as well as the enzyme ATPase, which provides the energy necessary for the transport of calcium ions into the

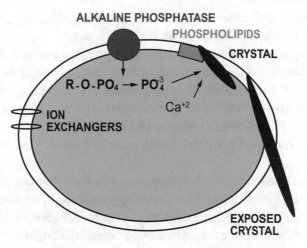

Figure 3.2: Mineralization of cartilage matrix is mediated by matrix vesicles, the membrane-bound, extracellular particles derived from chondrocytes. They possess the apparatus for concentrating calcium and phosphate and for nucleating mineral crystallization. Ultrastructural studies show that the crystals pierce the membrane and extend mineralization throughout the matrix.

vesicle against a concentration gradient. Matrix vesicles accumulate calcium and phosphate ions at their inner aspect [311]; the phosphate groups in the inner membrane of the matrix vesicles bind calcium, forming calcium-phosphate complexes [309]. The accumulation of calcium in matrix vesicles is aided by a family of calcium channel molecules known as annexins [312]. These complexes attract further calcium and phosphate ions that accumulate at the inner membrane and inside the matrix vesicles [313]. Thus, the calcium and phosphate ions form crystal nuclei at the inner membrane inside the matrix vesicle [314]. The concentration of proteoglycans is relatively increased within the calcifying cartilage of the hypertrophic zone of the growth plate [315], while the size of the proteoglycan aggregates diminishes between the growth plate cartilage and the metaphysis [316]. These changes in proteoglycan concentration and size are consistent with a mechanism where disaggregated proteoglycans contribute to matrix mineralization because of their ability to bind calcium [317] and aggregated proteoglycans inhibit calcium phosphate crystal formation [318]. The binding of type II and type X collagen to matrix vesicles depends on interactions with annexin V [312]. The association between type X collagen and matrix vesicles suggests that this collagen type also plays a role in the mineralization of the hypertrophic growth plate cartilage [319]. Thus, multiple mechanisms of mineralization occur.

The metaphysis lies between the last intact row of hypertrophic chondrocytes in the growth plate, and the diaphysis of the bone. In the bony metaphysis, the mineralized cartilaginous matrix of the hypertrophic zone is removed by lysosomal activity and vasculature invasion, new woven bone is formed by osteoblasts from the metaphysis (primary spongiosa), and cancellous trabeculae are remodeled by chondro and osteoclasts into lamellar bone (secondary spongiosa). The oxygen tension in the metaphysis is low (19.8±3.2 mm Hg) [297].

Surrounding the periphery of the growth plate is a population of cells in the shape of a wedge called the ossification groove of Ranvier that contribute to lateral growth of the physis. There are three cell types in the groove of Ranvier: osteoblast progenitors, chondrocyte precursors, and fibroblast-like cells [320]. The groove of Ranvier is, in turn, enveloped by a layer of fibrous tissue known as the perichondral ring of LaCroix, a collagenous network that provides additional mechanical support at the bone-cartilage junction of the growth plate by anchoring the periosteum firmly to the epiphyses [320].

Autocrine and paracrine factors such as hormones, enzymes, vitamins and growth factors affect the mineralization of the growth place by influencing chondrocyte proliferation, differentiation, biomolecule synthesis and/or intracellular calcium homeostasis. Because cells in different zones of the growth plate are in different stages of proliferation, differentiation, maturation and hypertrophy, different factors may act specifically on a particular zone.

Parathyroid hormone (PTH)-related peptide (PTHrP) and its receptor PTH/PTHrP receptor play an important, yet currently not well understood role in growth plate mineralization [321]. The PTH/PTHrP receptor is activated both by PTHrP and PTH. Perichondrial cells and proliferating chondrocytes in the periarticular region of the growth plate express PTHrP mRNA; whereas levels of expression of mRNA for the PTH/PTHrP receptor increase from the columnar proliferating zone chondrocytes towards the pre-hypertrophyc chondrocytes [322]. PTH and PTHrP regulate the expression of matrix metalloproteinases 2, 3, and 9 by growth plate chondrocytes [323]. Endochondral bone formation in the growth plates of MMP-9 null mice is disrupted [324]. PTH induces hypertrophic differentiation of growth plate chondrocytes, and vitamin D inhibits this effect [325, 326]. Indian hedgehog (Ihh) is also a major player in endochondral bone formation at the growth plate, regulating chondrocyte proliferation and differentiation, and osteoblast differentiation. Pre-hypertrophic and early hypertrophic chondrocytes in the growth plate secrete Ihh (reviewed by [292]). Ihh stimulates expression of PTHrP [327], and aided by PTHrP synthesis, Ihh inhibits the hypertrophic differentiation of chondrocytes and thus delays matrix mineralization [327]. A negative feedback loop results because Ihh/PTHrP interactions inhibit chondrocyte hypertrophic differentiation, which maintains the cells in a proliferative state thus inhibiting Ihh expression [321]. Through stimulation of PTHR synthesis, Ihh determines the distance from the end of the bone at which chondrocytes stop proliferating and become hypertrophic [292]. In developing bones, growth plate chondrocytes express Ihh, whereas chondrocytes in the ossifying epiphysis do not [328]. FGF and BMP signaling interact with the Ihh/PTHrP pathway during growth plate mineralization. Whereas FGF inhibits Ihh expression (independent of its stimulating activity on chondrocyte proliferation), BMPs act in a positive regulatory feedback loop with Ihh [329].

Systemic hormones also influence the growth plate. Growth hormone is produced by the pituitary, and it affects cellular proliferation in the growth plate by influencing the commitment of chondrocyte precursors into a proliferative state; it also stimulates osteoclastic resorption of primary spongiosa [330]. Growth hormone also stimulates insulin-like growth factor-I, which plays a crucial role in stimulating the proliferation and maturation of chondrocytes in the growth plate, while preventing apoptosis [331], and stimulating proteoglycan synthesis [332]. Calcitonin can accelerate growth plate calcification [333]. Glucocorticoids inhibit bone growth [334]. High-dose administration of corticosterone into a rat long bone model resulted in increased frequency of apoptosis in terminal hypertrophic chondrocytes, and reduction in the width of the cartilage growth plate, although both effects were reversible [335]. Vitamin D as the metabolites $1,25\text{-}(OH)_2D_3$ and $24,25\text{-}(OH)_2D_3$ regulates the differentiation of the mesenchymal cells into chondroblasts. Vitamin D metabolites influence cells from different zones of the growth plate differently: reserve zone chondrocytes respond primarily to $24,25\text{-}(OH)_2D_3$, while growth zone chondrocytes respond primarily to $1,25\text{-}(OH)_2D_3$ with increased fluidity of matrix vesicles and alkaline phosphatase activity [336, 337]. It was recently reported that vitamin D stimulates the actions of growth hormone and insulin-like growth factor as well as has a role in growth hormone's ability to reverse glucocorticoid inhibition of human growth plate chondrocytes [338]. The effects of retinoic acid, a vitamin A metabolite, on growth plate chondrocytes include inhibited proliferation of proliferative zone chondrocytes, decreased matrix synthesis, and decreased cell hypertrophy, with subsequent detrimental effects to mineralization [339]. Hypervitaminosis A results in premature closure of the epiphyseal growth plate [340]. Fibroblast-like-growth factor stimulates mitosis [341], as does platelet-derived growth factor [342], which also stimulates protein synthesis [343]. TGF-β controls cell replication and differentiation and inhibits the degradative cytokine IL-1.

3.3   ARTICULAR CARTILAGE

During early development and chondrogenesis, precursor mesenchymal cells differentiate into the cartilage lineage by a process that involves cell migration, aggregation, condensation and chondrocyte differentiation [344]. These cells form the cartilage anlage, which is a template of the future skeleton. In the first stages of formation, this appendicular skeleton is continuous, with no spaces or separations. As chondrification starts, changes occur in the regions of the future synovial joints. An "interzone" is created, which determines the location of the future joint and gives rise to its various components. The interzone contains three layers: two parallel chondrogenic layers covering the opposing surfaces of the future joint separated by a narrow, less dense intermediate layer. Simultaneous to the appearance of the interzone, the joint capsule begins to differentiate from the intermediate and deep mesoderm where the perichondrium and periosteum also rise. Blood vessels invade the evolving joint capsule and reach the blastemal synovium, but they do not enter the central region of the future joint. Intra-articular structures such as the menisci and the cruciate ligaments appear as cellular condensations in the intermediate layer. Once the articular cartilage is contoured from the two chondrogenic layers and the intra-articular structures are established, a multitude of very small spaces begin to appear in the intermediate zone, eventually coalescing to form the joint cavity. This process is called joint cavitation, and it is believed to be enzymatically mediated. The chondrogenic layers from the contours of the opposing joint surfaces and they take along the remaining undifferentiated cells from the intermediate layer to form synovium. Cavitation occurs so quickly that joint spaces are formed by the 10th week of gestation. At that point, because innervated muscle has developed, some motion can begin although *in utero* motion enhances joint development only to a small degree and nowhere near post-natal movement. The timing of interzone formation and joint cavitation is not the same for all the joints in the body. Small joints in the hand and feet show considerable delay between interzone formation and cavitation, and others such as those in the distal interphalangeal joint do not go through the homogenous three-layer zone [345].

Increased levels of hyaluronan facilitate the cell migration and aggregation that take place during the formation of the interzone. Wnt 14 is expressed early during the formation of the interzone, and it induces uridine diphophoglucose dehydrogenase and other joint development genes such as BMP-5 also known as cartilage-derived morphogenetic protein-1 or CDMP1. The chondrocytes in the interzone express the COL2A1 gene and differ from the chondrocytes in the growth plate in the sequence of gene expression. In the early stages of development, the epiphyseal growth plate is continuous with the future articular cartilage surface and indistinguishable from it. Even in many human joints at birth, the articular cartilage is not distinguishable from the growth plate. However, soon after birth a secondary ossification center appears in the epiphyseal cartilage and divides it into the metaphyseal growth plate and the articular cartilage. PTHrp and Ihh are expressed in a zone-specific manner in post-natal developing joints. In particular, Ihh is strongly present in chondrocytes at the articular surface, suggesting a role in prevention of hypertrophy, mineralization and ossification (reviewed by [346]).

CHAPTER 4

Responses to Injury and Grafting

4.1 BONE

4.1.1 MICROFRACTURES

Bone integrity may be damaged by trauma, infection, tumors, and genetic disorders. Like any material subject to cyclic loading, bone will show evidence of fatigue in *ex vivo* experiments, with the accumulation of microstructural cracks [347]. In 1960, Frost described the presence of small cracks with linear morphology in a series of human rib samples obtained at autopsy [348]. Those "native" microcracks were typically between 30 to 100 μm in length. Considerable evidence has accumulated to support Frost's idea that one of the consequences of bone remodeling is the removal of microcracks. A series of studies by Burr and coworkers showed a two- to three-fold increase bone microcracks and concomitant reduction in mechanical properties in the bones of dogs treated with long-term bisphosphonates to suppress remodeling [349]. Experimental evidence supports the view that microcracks initiate targeted remodeling [350] and that osteocyte death due to disruption of the vital canalicular network may be central to targeted bone remodeling [351]. In fact, the spatial and temporal associations of osteocyte apoptosis, bone microdamage, and intracortical remodeling support the view that osteocyte apoptosis is a key controlling step in the activation and/or targeting of osteoclastic resorption in response to bone fatigue. Experimental use of a pan-caspase inhibitor following fatigue loading suppressed osteocyte apoptosis and activation of osteoclastic remodeling around microcracks [352]. Although the molecular details of signaling between the cells are not yet fully clear, it is consistently observed that osteoclasts are more abundant in areas of microdamage, osteocytic apoptosis, and reduced vascularity; the process of bone remodeling is essential for renewal of bone tissue. Another benefit of continuous bone tissue remodeling is the ability of bone to regenerate and heal fractures. Most other tissues lack regenerative capacity and heal by scar formation, considered a default pathway.

4.1.2 BONE FRACTURES

Fractures disrupt cellular viability, vascular connections, and structural integrity of bone. They occur when the load to a bone exceeds the range to which it has adapted. Fractures are often classified according to the force that causes them. They may be low-energy fractures, as in a torsional fracture, high-energy as in those caused by automobile accidents, or very high-energy as in those caused by gunshots. Fragility fractures are those that occur with low energy, such as a fall from standing, to bones weakened with osteoporosis. Bone tumors can cause pathological fractures in areas weakened by a primary or metastatic lesion. Fractures may be caused by a load that is higher than the ultimate strength of bone, or by repetitive application of loads of lower magnitude, as in the case of fatigue fractures. Fatigue fractures may result from the cyclic application of a small load over a long period of time, or a few repetitions of a high load [353]. Stress fractures occur when low magnitude forces result in progressive accumulation of micro-damage and failure. They are common in military recruits undergoing basic training. Bone may sometimes eventually fail through fracture-crack propagation. Another way to classify bone fractures is by the mechanism of injury. Direct force fractures occur when forces are applied directly at the fracture site. Indirect force fractures result from forces being transmitted from distance away from the fracture site. An example of an indirect force fracture is one that occurs when there is a strong muscle contraction across a joint with a fixed distal segment such as the patella. The classification of fractures according to the direction of the fracture lines includes: transverse (fracture running perpendicular to the bone), oblique (fracture runs across the bone at an angle of 45 to 60 degrees), spiral (fracture has a torsional component), comminuted (fracture has more than two fragments), and impacted (fracture ends are compressed together). Additionally, according to the degree of associated soft tissue injury, fractures may be simple (no associated skin injury), compounded (the overlying skin if broken), complicated (there is damage to the muscular, neurovascular, ligamentous, and/or visceral components, or the fracture is intra-articular), or uncomplicated (the fracture has only a minimal amount of soft tissue injury) [354].

The direction of forces that fracture bone may be such that the resulting stresses are compressive, shear, or tensile, alone or in combination. Compressive loads typically cause oblique fractures, whereas torsional stresses produce spiral fractures. Tensile stresses cause flat fractures, but these are less common. Long bones fail in patterns that follow some basic rules for all types of stress. For example, cortical bone is generally weak in tension and shear, in particular along its longitudinal plane. It often fails in that manner [132].

4.1.3 FRACTURE HEALING

Bone tissue is capable of true regeneration in response to injury. Repair is the process of response to an injury, usually as a scar, whereas regeneration is the restoration of the injured parts with characteristics indistinguishable from the original material. Postnatal skin is not capable of regeneration, but repairs quickly by scar formation. Scar is a poorly organized tissue that can be regarded as the quick way to seal the skin and avoid risk of invasion and infection.

In order for proper fracture healing to occur, there must be enough vascularity and viable cells available to mediate each of the phases of repair. Optimally, a stable mechanical environment must exist, and if not, it can be achieved with casting or surgical screws, plates, or rods. Blood supply is disrupted to the fracture site, causing tissue necrosis for a distance from the fracture. Depending on mechanics of the injured site, bone fractures heal by one of two mechanisms: primary or secondary.

Primary healing is that which takes place spontaneously in fractures that are partial or stable and that are aligned so that their surfaces are very closely apposed. If immobilization can be achieved by external fixation and/or reduction to good anatomic position, these types of fractures do not often require surgical intervention. External or internal fixation may achieve the stability required for good healing. Fractures treated by successful operative primary internal fixation show evidence of primary bone healing, without cartilage or fibrous tissue formation during the process of healing [355]. In primary bone healing, osteons close to the fracture site are devitalized due to the disruption in the local blood supply. Osteoclasts form cutting cones at the ends of the Haversian canals, and advance at a rate of 50 to 80 μm per 24 hours, resorbing the bone around the fracture site and leading the capillary extension across the fracture. Osteoblasts trail behind the osteoclasts and deposit new woven bone, thus creating osteons that connect both sides of the fracture. Reduction that obliterates gaps between the fracture fragments and immobilization by fixation of the fractured segments is indispensable for primary healing. Reduction is achieved non-invasively with weight-bearing casts or with external fixators, or operatively with internal fixators such as plates and screws. With internal fixation, it may take 5 to 6 weeks for new bone to fuse the ends of the fracture [356].

Secondary bone healing is more common than primary healing, and it involves a sequence of events that also result in regeneration of the injured bone. The secondary healing process consists of four stages: inflammation, soft callus formation, hard callus formation, and remodeling.

The inflammation stage begins immediately after the fracture and persists until the initiation of cartilage or bone formation. It usually lasts 3 or 4 days, and it ends with a decrease in pain and swelling. The inflammation stage involves the formation of a hematoma and a fibrin clot and the arrival of platelets, polymorphonuclear neutrophils, and monocytes or macrophages. The fibrin clot brings with it invasion by fibroblasts, mesenchymal cells, and osteoprogenitor cells. The bone at the edges of the fracture site becomes necrotic due to the disruption in vascularity; osteocytic autolysis by the release of lysosomal enzymes help to degrade the tissue [356]. Local and infiltrating macrophages phagocytose debris and osteoclasts resorb fragments and foci of necrotic cortical and trabecular bone.

A very active stage, known as the soft callus, follows. Osteoprogenitor cells in the cambium layer of the periosteum proliferate rapidly during the soft callus stage, and they differentiate into chondrocytes and osteoblasts [357]. The conditions at the fracture site are relatively hypoxic during the soft callus stage [358]. The periosteal progenitor cells form a callus of cartilage externally and internally, which serves to immobilize the fracture segments such that the bone may be loaded long before the process of union is complete [359]. The soft callus may also be viewed as a temporary tissue that provides a relatively immobile scaffold for the orderly formation of vascularized bone. Soft callus formation lasts for 3 to 4 weeks until the fractured fragments are connected through repair fibrous or cartilaginous tissue and can no longer move freely [6].

In the hard callus stage, the soft callus becomes extensively vascularized and is replaced by woven bone. This occurs by a combination of endochondral bone formation and some areas of intramembranous bone formation. Intramembranous bone formation takes place in the more vascularized areas, with mesenchymal cells or preosteoblasts such as those in the periosteal callus forming bone directly without a cartilage stage. The hard callus stage is associated with a marked increase in vascularity and in cellularity. The conditions at the fracture site remain relatively hypoxic. The osteoclasts continue to actively remove the remaining dead bone. The hard callus stage takes 3 to 4 months in major long bones, and significantly less time in pediatric patients [6].

The final stage of fracture healing is the remodeling stage, wherein woven bone is gradually reorganized into lamellar bone, and the medullary canal is reconstituted. Osteoclasts remodel the external bone surface and remove the hard callus. The oxygen supply returns to normal. Remodeling may take years to complete, and it occurs much faster in pediatric patients than in adults [6].

It can be said that angiogenesis leads the way for ossification. Vascularity at the fracture site undergoes important changes during the stages of fracture healing. These changes allow for proper bone repair/regeneration to take place. Sumner-Smith reviewed the timeline of changes in vascularity at the fracture site [360]. For fractures with minimal displacement where the blood supply is disturbed, there is an initial decrease in blood flow due. Within hours to a few days, there is a significant increase in vascularity that peaks at approximately 2 weeks under conditions of minimal displacement and good stabilization. Thereafter, as healing progresses, vascularity gradually returns to normal. Increased vascularization at the fracture site has been reported for juvenile mice compared with middle-aged and elderly ones [361].

4.1.4   DISTRACTION OSTEOGENESIS

Distraction osteogenesis (DO) is the technique of expanding bone volume by gradual distraction across an osteotomy or corticotomy with a device that ensures enough stability to allow angiogenesis and new bone formation while also expanding adjacent soft tissues (Figure 4.1). It can be viewed as a controlled series of microfractures in which there is a prolongation of the proliferation and activity of osteoblasts, accompanied by angiogenesis. It is absolutely dependent on the regenerative capacity of bone tissue and harnessing that potential in a temporo-spatial manner. DO has been used effectively to augment bone in patients with skeletal deformities such as mandibular defects [362], craniofacial deformities [363], and alveolar bone defects [364].

The technique was first described in 1905 [365] and was refined by the Russian surgeon Ilizarov through comprehensive laboratory studies [366, 367]. Many DO models have been developed for experimentation in mice [368], rats [369], rabbits [370], dogs [371], and minipigs [372].

4.1.5   COMPARISON OF POST-NATAL FRACTURE HEALING AND
SKELETAL EMBRYOGENESIS

Embryonic bone development and growth are complex processes that result in the orderly modeling and expansion of the diverse shapes of skeletal structures. Modeling is defined as process of sculpting of developing bones into the requisite shape. Mechanical factors such as those applied to cranial bones that result from expansion of the embryonic brain, for example, contribute to the modeling of the skull bones. The human neurocranium forms from a mesenchymal capsule around developing neural tissue

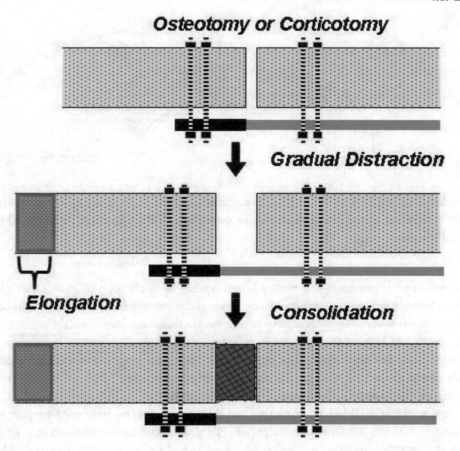

Figure 4.1: Distraction Osteogenesis. A distraction device is secured to bone with screws across a surgical osteotomy (cut through the bone) or corticotomy (through the cortex). A mechanism is activated at intervals to gradually separate the two sides and to amplify and prolong fracture healing in the gap. After the desired amount of elongation has been achieved, the device ensures fixation and maturation of bone in the gap.

and is divided into the membranous calvaria and the cartilaginous cranial base [373]. Ossification centers appear by 6 to 7 intrauterine weeks and by birth the neurocranium has achieved 63% of its ultimate size. This is in marked contrast with the facial bones that have achieved only 40 to 45% of adult size. To expand the growing skull bones, changes in size and curvature (shape) are accomplished by osteoblastic bone deposition on the ectocranial surface and osteoclastic bone resorption on the endocranial surface (Figure 4.2). Bone apposition occurs on one surface and resorption on the other. Such coordination of osteoblast and osteoclast activities in space is different from the coupling of

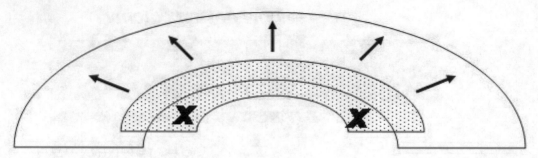

Figure 4.2: Modeling of a curved cranial bone (stippled area) requires expansion of the ectocranial surface by osteoblasts (arrows) and resorption by osteoclasts of the endocranial surface (X) to achieve growth and contour.

bone resorption and formation that occurs in remodeling of adult bone (Figure 2.8). Other major distinctions include the important roles of osteocytes and angiogenesis in directing bone remodeling.

Bone tissue is striking in its ability to regenerate following injury, unlike other tissues that heal with scar tissue. It is common dogma in elementary textbooks that fracture healing recapitulates embryonic skeletal development. In its simplest form, the idea is proposed that the same cellular and molecular mechanisms that mediate the development of embryonic bones are recalled into action upon injury to the skeleton. Cellular and molecular markers of chondrogenesis and osteogenesis that are expressed during development have been shown to appear during fracture repair, leading to the hypothesis that the regulation of embryonic cell differentiation is conserved in post-natal life. Although it is surely important to catalog all the cells and factors involved in the developmental stages of membranous and endochondral ossification, it appears overly simplistic to equate post-natal fracture healing and embryonic skeletogenesis. It is widely anticipated that knowledge about molecular mechanisms governing skeletal patterning and morphogenesis should support the development of new approaches to regenerate skeletal tissue in patients. We should be cautious about critical differences that are not often acknowledged and that may limit the therapeutic applicability of the details learned about mechanisms in skeletal development.

Classical studies of comparative embryonic skeletogenesis describe common events of patterning, cell differentiation, and morphogenesis that give rise to the cartilage and bone of the skeleton. More recent research, particularly investigations with mouse and Drosophila genetics, has identified critical genes that regulate skeletal development, such as Runx2 [374], the hedgehog genes Hedgehog signaling in skeletal development, PTHrP [375], the Wnt genes [376], and members of the TGF/BMP superfamily [377].

Studies that show that a particular gene is activated early in skeletal development often conclude that the gene must play a key role in bone formation. That inference can be further supported if disruption of the gene results in a skeletal derangement, even if only in certain bones. Studies evaluating fracture healing in post-natal mutant mice reveal the difficulties in interpreting the observations because of the complexity of tissues and factors involved in callus formation and remodeling [378]. It can be a challenge for investigators to draw conclusions about the changes initiated by the absence of a single gene unless alterations are quantified for an often impractical assortment of outcome measures. Alterations may be direct or indirectly dependent on the gene of interest. Interruption of the action of a single gene can alter multiple processes, as demonstrated by exemplary studies of fracture healing in mice lacking TNF-α signaling [379]. Such careful investigations avoided the risks of falsely concluding, for example, that loss of a gene enhances repair when it in fact inhibits the resorption of early, immature repair tissue.

It is clear that many genes are involved in both development and post-natal fracture healing, particularly those that mediate the endochondral sequence. For example, homozygous mice with a null mutation in matrix metalloproteinase-9 (MMP-9) showed delays in vascularization of growth plates and in apoptosis of hypertrophic chondrocytes [324]. A subsequent study of post-natal, non-stabilized fracture healing demonstrated a transient delay in vascularization of and removal of cartilage callus in the mutant mice, compared with wild-type littermates [380]. Thus, there was parallelism in effects of the loss of MMP-9 on skeletal development and post-natal fracture healing. Those and other reports with mutant mice describe differences between skeletal development and fracture healing. Mice with a null mutation in MMP-9 showed no developmental abnormalities in bones that were formed by intramembranous ossification, but in a model of primary fracture healing, they did show a retardation in post-natal intramembranous bone formation and, in fact, unexpected formation of cartilage [380]. Another example of differences between skeletal development and post-natal fracture healing is provided by studies with mice lacking TNF-α signaling. First, mice lacking TNF-α display no skeletal abnormalities [381]. Second, mice deficient in the two forms of TNF-α receptors, p55 and p75, are also overtly normal [382]. Yet, the mice lacking the TNF-α receptors were shown to have impaired post-natal fracture healing, compared with wild-type mice [379]. The results argue against an obvious role for these receptors, and hence TNF, in normal mouse skeletal development and growth, but that TNF-α is essential post-natally for proper recruitment and differentiation of chondrocytes, and also for osteoclast recruitment and action in experimental fracture healing.

Elegant studies with mice engineered to lack BMP-2 in limb bud mesenchyme showed that they displayed proper skeletal patterning, early limb chondrogenesis, and the normal transition from cartilage to bone that occurs before birth [383], but that they failed to heal experimental or spontaneous post-natal fractures [384]. These investigators concluded that BMP2 expression may not be as critical for limb patterning as previously believed, but that it is absolutely necessary for the signaling cascade that governs fracture repair.

Fetal wound-healing studies offer insights into the frequent discordance observed between fetal skeletal development and post-natal osteogenesis. Skin wounds made in early-gestation fetuses have been shown to heal without inflammation or scarring, a phenomenon attributable to the abundance of hyaluronic acid in the wound matrix, local deficiency in acute inflammation, and relative deficiency of TGF-β, PDGF, and hyaluronic acid degradation products [385]. Fetal skin wounds heal without scar before a certain gestational age, after which point typical scar formation occurs. In mice, scarring of wounds occurs after embryonic day 18.5 (average gestation period for mice is 20 days) [386]. It is very important to define for each species the age for the conversion and what else is changing at that time. Some discrepancies in the literature may be attributable to gestational age of the conversion and even to differences in different locations. It is notable that in the same animals that show scarless repair of skin wounds, incisions in the diaphragm repair with abundant scar tissue [387]. Research with different animal models indicates that the extent of inflammation and mediators at the wound site influence scar formation [387]. One of the major explanations for fetal scarless healing is that early in gestation, when fetal skin is rapidly growing, its ECM is a hyaluronan-rich, loose network relatively lacking in TGF-β and its pro-fibrotic effects. Wounding this unique tissue results in a scarless wound also composed of fine reticular collagen in a hyaluronan-rich matrix [388]. The sustained deposition of hyaluronan is unique to fetal skin wounds and has been attributed to the fetal paucity of TGF-β [389]. Comparison between postnatal and fetal wound healing reveals important differences in roles for inflammatory response, cellular mediators, cytokines, growth factors, and ECM modulators. Much research is motivated by the goal to reduce excess scarring in patients. The idea that the absence of TGF-β and inflammation in fetal skin wound healing is the key to scarless healing found support from wound healing experiments in animal models [390]. There is an obvious advantage for a post-natal organism to be able to rapidly seal off open wounds; thus, dermal scar tissue, although not a regeneration of normal organized skin structure, serves to protect against infections by walling off the lesion very quickly. In spite of the great interest in applying knowledge from fetal skin wounds to reducing scarring in human patients, especially to reduce fibrosis that is attributable to post-natal abundance of TGF-β, there are currently no scar reduction strategies based upon embryonic healing that are available for clinical use [391].

   In contrast, very little information is available about the biology of healing with experimental fetal osseous wounds, but given the ability of adult bone to regenerate itself, it may be anticipated that injuries to fetal bone would heal well. One study with a fetal lamb model indicated that ulnar healing appeared complete within 21 to 40 days [392]. We conducted a pilot study in fetal rabbits that showed unexpected results. Parietal bone defects measuring 4-mm × 5-mm were made in rabbits at 24 fetal days of age, and harvested calvariae were examined histologically at weekly intervals. The average gestational age for rabbits is 31 days; thus 24 fetal days corresponds to mid-second trimester. In specimens harvested at 1 week after wounding (at birth), there was no evidence of inflammatory cells, angiogenesis, or bone formation; the defects were devoid of any tissues. In specimens harvested at 7 days after birth, signs of spotty bone regeneration were evident [393]. A material know to be inflammatory in adults, polyethylene particles, was placed into parietal wounds in another group of fetal rabbits; this elicited a mild inflammatory infiltrate and angiogenic response within 3 days and resulted in osseous wound repair *in utero*. The expectation had been that empty wounds would heal spontaneously and that inflammatory particles would inhibit bone formation, as has been observed in adult animal models [394]. These surprising results shed new light onto the importance of local inflammatory cells on bone healing that are not part of embryonic bone formation. The absence of bone formation in empty parietal wounds can be explained on the basis of cranial development in rabbits. Serial analysis of parietal bones showed that the bone is a thin structure devoid of vessels, marrow spaces, and myelopoiesis until 7 days after birth, a time coincident with the initiation of bone formation in the empty wounds. In contrast to the post-natal development of calvarial diploë and myelopoiesis, tibial myelopoiesis was vast in 27 day fetuses; this shows asynchronous site-specific development of hematopoietic marrow. The data suggest that the local absence of myelopoiesis may render fetal calvariae incapable of the inflammatory phase of regeneration. The polyethylene particles were able to recruit a small number of inflammatory cells and their mediators, coincident with bone formation. Thus, fetal responses to osseous wounding and to inflammatory implants were exactly opposite to that seen in adults and may be associated with the late development of inflammatory cells. These findings support the view that the dissimilarities between skeletogenesis and post-natal fracture healing may be more consequential than their similarities. More knowledge about the differences between modeling and remodeling and the differences between skeletal embryogenesis and fracture healing is required for advances in skeletal tissue repair.

4.1.6 BONE GRAFTS, IMPLANTS, AND BONE SUBSTITUTE MATERIALS

Although fractures have the capacity to heal with true regeneration of bone tissue, in some cases, the result may not be ideal because of systemic problems such as poor nutrition, comorbidities such as diabetes, and use of medications such as glucocorticoids. Poor results can occur because of local problems such as comminution into many fragments, infection, inadequate immobilization, or anatomical problems. Data from 2004 show that nearly 15.3 million fractures in the United States were treated in physician offices, emergency and outpatient clinics, and hospitals [395]. Fractures in certain bones, notably the tibia, femur, and humerus, are at risk for non-union or delayed union. The tibia is the most common problem site, with delayed healing or nonunion occurring in 13%, in a typical prospective study [396]. Risks for tibial union include open rather than closed fracture, fractures with gap greater than 5 cm, and smoking history. Tibial non-unions may show excess callus due to mobility and are called hypertrophic non-unions. On the other hand, they may be classified as atrophic non-unions because of an absence of callus and poor vascularity, sometimes with the appearance of a joint space (tibial pseudoarthrosis).

Bone grafts are used to promote bone formation in osseous voids from congenital or acquired conditions. The biological host response to a bone graft depends on the type of graft. Bone grafts differ on their source (autogenous, allogeneic, xenogeneic), method of storage (fresh, frozen, lyophilized), and structural form (cortical, cancellous). Considered the "gold standard" to replace bone in discontinuity defects or cystic cavities or to augment deficient bony contour, fresh autogenous graft is a portion of bone harvested from one part of the body and transplanted to the recipient site. Advantages of autogenous bone include early vascularization, excellent incorporation of the graft, safety from disease transmission, and absence of an immune rejection response. Common sites from which the graft is harvested are iliac crest, tibia, fibula, and scapula or, in the case of craniomaxillo-facial reconstruction, the symphysis, maxillary tuberosity, mandibular retro-molar area, or zygoma. Cancellous autografts are readily revascularized, which supports the viability of transplanted cells. Remodeling and reorientation of the trabeculae to a mature pattern follow. In the case of fresh autogenous cortical bone graphs, very few cells survive transplantation [397]. Osteoconduction may involve "creeping substitution," a term used to describe the resorption of the scaffold as new bone forms. Revascularization of cortical autografts is much slower than in cancellous autografts, and it entails osteoclastic resorption and osteoconductive new bone formation. It is a clinical failure when resorption of the implant exceeds the volume of new bone formation. Typically, cortical autografts are not fully substituted with viable new bone, but they can be useful and maintain volume because of the endurance of the cortical bone [398]. Although remote from the osseous defect and thus requiring a second operative site, the iliac crest is favored because it is accessible and rich with os-teoblast progenitor cells and harvesting it does not compromise the structure of the ilium. The major mechanism of healing is direct osteogenesis by the osteoblasts and preosteoblasts that survive the harvesting and transplantation procedure. We classified three basic mechanism of bone formation in response to different types of bone grafts/implants (Table 4.1), but for many grafts and implants, admixtures of mechanisms are involved [399]. The harvesting of fresh autogenous bone adds to op-erative time, and the quantity of available bone may not be sufficient, especially for infants and frail elders [400]. Thus, surgeons and researchers have sought to develop alternatives to fresh autogenous graft for skeletal reconstruction.

Table 4.1: Mechanisms of Bone Healing

Type	Physiologic Principle	Examples
Osteogenesis	Transplantation of viable os-teoblasts or pre-osteoblasts	Cancellous bone and/or marrow, Periosteum, Vascularized grafts
Osteoconduction	Expansion of bone from the mar-gins of the defect over or into a scaffold that can be resorbable or permanent	Cortical bone, Allogeneic banked bone, Synthetic materials
Osteoinduction	Phenotypic conversion of mes-enchymal cells into skeletal cells	Demineralized bone and dentin

Allogenic bone and other alternatives have been used in such settings as well as for situations in which the small size of a defect does not justify harvesting of the patients' own bone. An example would be in periodontal defects. Allogeneic bone from cadavers needs to be thoroughly cleansed and neutralized (de-antigenized) by freezing or lyophilization in order to be used safely in patients. Allogeneic bones and their derivatives have low risks of disease transmission and immunogenicity due to improvements in strict donor selection criteria and tissue processing. They are devitalized, cleaned, and treated to reduce inflammatory and immune reactions. Although they contain no live cells, they provide for load-bearing and osteoconduction (Table 4.1). Technically, the term "graft" should refer to viable tissue or organs, but the term is used as jargon, for example, as "allograft." Upon implantation, allogeneic bone is vascularized, and the matrix is resorbed by osteoclasts and replaced with new viable bone by osteoconduction [401]. The processing and effectiveness of allogeneic bone, however, exhibits some significant variability [402]. While cancellous allogeneic bone is usually incorporated completely, cortical allogeneic bone, like autogenous cortical bone, often remain as a mixture of viable new bone and old bone for a prolonged time (reviewed in [403]).

Osteoinduction is the conversion of non-osseous tissue into bone by an appropriate stimulus (Table 4.1). Demineralized bone was shown to be such a stimulus [404] and that it did so by an endochondral process [405]. It has been used in an array of clinical applications, including craniofacial [406], oral surgical [407], periodontal [408], hand reconstructive [409], and orthopedic procedures [410].

Other materials are used as bone substitutes, including minerals, ceramics, polymers, and metals. Depending on the host response to a test implant, materials can be defined as the following: (i) *toxic*, when the surrounding host tissue is injured; (ii) *nontoxic and biologically inactive*, in which a fibrous capsule forms around the implant; (iii) *non-toxic and biologically active*, in which there is formation of an interfacial bond between the host and the materials; and (iv) *non-toxic and resorbable*, in which the material eventually dissolves and the surrounding tissue replaces it [411]. Some implants such as poly(α-hydroxy esters) are resorbable, degrading upon implantation at rates that depend on the material's chemistry and architecture, but that ideally match the rates of new bone formation to preserve mechanical stability. Other implant materials such as metals are permanent. Even calcium phosphates are available as resorbable or non-resorbable forms, depending upon their crystalline density [412]. Different animal assays can be used to test for general biocompatibility and specific osteocompatibility [394].

In spite of their ease of availability, synthetic implants are not ideal. Their drawbacks include poor integration and inflammatory or immune reaction [413]. If they are permanent materials, they can be the nidus for infections [414]. These drawbacks promote the gradual resorption of bone at the material interface. The incorporation of synthetic grafts into the host bone constitutes a response to bone injury, and it involves inflammation and bone regeneration [415]. The utility of an implant depends on the ability of the material to not emit products that elicit immune/inflammatory responses, interfere with cellular activity, or interfere with the stroma required for the regeneration of tissue at the defect site [416, 417, 418]. Motion of the implant, with respect to the bone, may lead to failure of the bone/implant interface. Motion disrupts the regenerating osseous stroma and promotes the formation of a fibrous capsule with associated recruitment of inflammatory macrophages and wear/corrosion of the implant. As the implant/bone interface wears, motion with respect to the bone increases; this vicious cycle may lead to implant failure. Motion at the implant/host bone interface may be prevented by fixing the implant to the bone by mechanical or chemical means. Mechanical bonding is achieved by interdigitation of bone with bone cement or with irregularities in the implant's surface topography or by promotion of bone ingrowth into suitable porous surfaces. Chemical bonding is achieved by adsorption/bonding of proteins to the surfaces of implants, to which bone cells can attach. Whether mechanical, chemical, or a mixture of both, bonding limits the motion between the implant and bone, and it imparts a more favorable distribution of stresses to the surrounding bone [132].

In some cases, synthetic implants do not integrate to bone and instead become surrounded by fibrous tissue. This occurs when a fibrous capsule forms rather than bony ingrowth. Macrophages are often found at the surfaces of the material. A fibrous capsule can lead to failure of the implant and to restrictions in nutrient supply [419]. Medical status, especially nutritional [420], diseases of bone metabolism (osteoporosis, osteomalacia, hyperparathyroidism), rheumatic and hormonal diseases (rheumatoid arthritis and diabetes), immunological disorders and malabsorption syndromes can interfere with incorporation of synthetic implants [421]. Other factors such as smoking, bone quality, irradiation therapy, and bacterial contamination have been associated with failure of oral implants [422]. Particulate debris results when implant components move in relation to fragments of bone cement or against the surrounding bone, or when there is wear at the articulating surfaces of total joint replacement prostheses. Particulate debris induces changes in the implant environment and generates a peri-implant pseudo-synovium [423]. The debris may be metallic or polymeric, and depending on its size, quantity, chemistry, topography and shape of the pieces it may elicit different adverse events. Rough or sharp particles produce the most irritation to the surrounding tissue. Wear particles less than 5 mm in average dimension stimulate macrophages to produce osteolytic cytokines [424]. Particles of both polymethylmethacrylate cement and polyethylene of the articulating component of a prosthesis elicit foreign body giant cells [425]. Metal particles are often smaller than plastic ones and although they do not elicit an inflammatory response, they accumulate in cells and may cause mutagenic damage [426]. The particles can act as third bodies, and damage both the implant and the host bone surfaces [427].

4.2   CARTILAGE

Athanasiou and colleagues extensively describe types of cartilage injury, causes, and associated costs [428]. Although endowed with remarkable mechanical properties to transmit load, absorb impact, and sustain shearing forces, articular cartilage is limited in its ability to restore its structure or function upon injury or degeneration. This has been attributed to the absence of blood vessels and of regenerative cells that can repair defects of significant size. Articular cartilage lacks endogenous undifferentiated mesenchymal cells, and without blood vessels, new cells do not enter the injury site to aid the healing process. In aged individuals, healing is even more compromised because of the lower density of chondrocytes in the tissue, and the lower quality of ECM that they produce [429]. A type I, or partial-thickness, defect is one that is restricted to the substance of articular cartilage, and a type II defect is the result of an injury that penetrates and extends into subchondral bone.

Injury to articular cartilage occurs through loss of ECM macromolecules (for example, in osteoarthritis) or through mechanical trauma. In general, the smaller the defect, the better the self-healing ability of articular cartilage [430]. If the structural matrix is not significantly disrupted, articular chondrocytes attempt to repair the loss of ECM macromolecules by proliferating and increasing their levels of ECM biosynthesis [431]. If the injury is localized and superficial, for example an injury that frays the surface of articular cartilage, chondrocytes immediately at the edge of the injury die, and adjacent cells in the most superficial zone proliferate and form clusters and increase their matrix synthesis [432]. Nevertheless, repair is incomplete because the matrix secreted is not sufficient to fill the defect. Some isolated partial-thickness injuries that do not show significant healing over time, remain stable, and do not progress into degenerative osteoarthritis [432].

Articular cartilage injuries that penetrate the tidemark and extend into marrow in the subchondral region, elicit the migration of undifferentiated cells and vessels to the defect site [433]. This type of injury is called an osteochondral defect. Healing of large lesions involves formation of a fibrin clot and ultimately immature cartilage or fibrocartilage. The osseous portion of the injury is reconstituted, and the subchondral plate is restored. The immature cartilage contains disproportionate amounts of type I collagen, insufficient amounts of type II collagen and large aggregating proteoglycans, as well as other anomalies with respect to healthy articular cartilage. Its mechanical function is poor, and it results in the formation of an arthritic defect in the surface of the tissue [46].

In some settings, there can be transient repair of cartilage. The repair of a small osteochondral defect repair starts with the formation of a fibrin clot. The clot may fill a defect and be retained there as a scaffold for repair if the lesion is only a few millimeters wide. Formation of the clot is followed by invasion of inflammatory cells, fibroblasts that begin to deposit a collagenous matrix, and mesenchymal cells which develop a round shape and secrete a cartilage-like matrix. Regenerative cells may originate in synovium or the superficial layer of the articular surface. Six weeks after injury, the chondral portion of the defect is cartilage-like, containing chondrocyte-like cells, collagen type II, proteoglycans and some collagen type I, while the bone portion of the defect contains immature bone, fibrous tissue and cartilage. Six months after injury, subchondral bone is fully repaired with a matrix that contains mostly bone but also some fibrous and some cartilaginous tissue. The chondral aspect of the defect may be partially repaired, or it may contain substantial amounts of fibrous tissue. Its composition is intermediate between hyaline cartilage and fibrocartilage. Unfortunately, with the passage of years, in most individuals the repair cartilage tissue becomes progressively more fibrillar and the cells become fibroblastic. The tissue fibrillates and fragments, exposing the bone surface. The joint becomes stiff and painful; this condition is called post-traumatic osteoarthritis [434].

Injuries perpendicular to the articular surface disrupt the ECM and kill the chondrocytes at the edges of the injury; vertical fissures that deepen with time are common. Proteoglycans at the surface zone are released through these fissures. Although adjacent chondrocytes may proliferate, form clusters, and synthesize increased amounts of matrix components, they do not migrate to the injury site and therefore do not fill the fissures. Because the defects are left unfilled, the adjacent tissue deteriorates with wear such that fragments of cartilage may tear loose and get released into the joint space. Eventually, the subchondral bone is exposed and eburnation occurs [5, 432, 434].

Pain from articular cartilage injury is derived from contact of the subchondral bone plate and from bare bone surface exposure. Further, breakdown products from articular cartilage degeneration may cause joint effusions, capsular distension, synovitis and other mechanisms of pain. As the subchondral bone plate hardens, secondary vascular venous congestion in the medullary cavity also causes pain [5].

The short and long term goals of cartilage repair are to alleviate symptoms and to prevent progressive breakdown of the cartilage surface and development of osteoarthritis. It is necessary to generate a tissue that fills the defect and integrates with the surrounding articular cartilage and the underlying subchondral bone plate. The tissue must have similar viscoelastic mechanical properties as healthy articular cartilage and be able to maintain itself [5]. This can be approached with procedures such as microfracture, osteoarticular autograft transfer system (OATS), mosaicplasty, and autologous chondrocyte expansion and implantation. With current therapeutic alternatives, the repair tissue that forms after clinical intervention may be fibrous, transitional, fibrocartilage, hyaline cartilage, articular cartilage, bone, or a mixture of any of these [435]. The predominant type of repair tissue determines the long term outcome of the joint, with hyaline or articular cartilage having the best prognosis, but being the most challenging to achieve.

The synovial joint spaces containing cluster of cells for bone, ligament, capsule, and cartilage form within 7 or 8 weeks of human embryonic development [436]. Most of the immature cartilage that forms during skeletal embryogenesis is temporary and gets replaced by bone during epiphyseal development. However, those regions closest to the synovial cavity persist past fetal development, and they become the permanent articular cartilage of the adult [5].

Immature articular cartilage contains abundant chondrocytes that actively proliferate and secrete ECM in both in the superficial and deep zones. As the individual approaches skeletal maturity, chondrocyte proliferation and ECM secretion become more limited to the deep zone [3].

CHAPTER 5

Clinical Applications for Skeletal Tissue Engineering

This chapter describes common diseases of bone and joints that often require reconstruction. The aim is to introduce new investigators to the many clinical settings for which tissue engineering may provide improved outcomes. New tissue engineering technologies raise expectations for tissue regeneration. Understanding of the pathophysiology underlying different skeletal disorders is necessary in order to design proper treatments, whether they are medical, surgical, or innovative tissue engineering approaches. Attention must be given to the genetic, anatomical, vascular, and systemic environment of a skeletal defect when planning treatment. It is standard practice for orthopedic surgeons to prepare recipient graft sites to promote remodeling and restoration of functional tissue, for example, by ensuring exposure of bleeding bone, but more extensive manipulations may be needed for systemic or genetic diseases. Local and systemic disorders may need to be managed with conventional therapy before use of an engineered device. It is obviously imprudent to place an engineered device into a site that is infected. Similarly, avascular, necrotic, or irradiated sites are unlikely to provide an environment favorable for regeneration. The first principles of trauma surgery are to ensure closure, sterility, and stability of a fracture site. Likewise, nutritional insufficiency would need to be corrected before intervention with a device. Comorbidities and medications being taken by a patient can have a profound effect on response to orthopedic surgery. It will be evident that some tissue engineering approaches may be unsuitable or may require extensive modifications or further refinements for application to certain disorders.

5.1 DISORDERS IN BONE

5.1.1 FOCAL DEFICIENCIES

5.1.1.1 Developmental

Craniofacial Deformities. Craniofacial deformities can be caused by combinations of polymorphic genes, nonspecific environmental agents, deficiencies in factors such as vitamin B folic acid, and some anti-seizure medications taken during pregnancy [437]. The most common deformities that include skeletal voids are cleft palates, other oral facial clefts, and microsomia [438]. Babies and children with oral clefts have difficulties with feeding, speech, and development of dentition as well as frequent ear infections and hearing loss. Bone grafts are commonly used to reconstruct normal anatomy. The question is when to intervene surgically. With advances in anesthesia and bone grafting techniques, the view evolved that early operative correction of craniofacial deformities in a child would provide a more normal functional matrix within which the craniofacial skeleton would complete its growth. In addition to optimizing growth by building structures in a more normal anatomic position, early correction could minimize secondary deformity and improve the child's appearance, body image, and socialization. For example, the natural history of hemifacial microsomia shows progression of facial asymmetry as the jaw becomes progressively more asymmetric with growth on the contralateral side outpacing the affected side [439]. There may be a crossbite on the affected side and ultimately muscular hypoplasia matches the degree of bony abnormality. Thus, to avoid the progression, intervention is recommended when the child is in the mixed dentition stage. The rationale takes advantage of the extensive potential for vertical growth of the midface as the deciduous teeth are shed and the permanent teeth erupt. Follow-up studies show endurance of symmetry in patients whose treatments were initiated early [440].

For other parts of the anatomy, however, well-documented data show that impairment of growth can occur after intervening too early. On the basis of evidence that an early operation for midface hypoplasia or orbital hypertelorism may interfere with anterior facial growth, for example, it is recommended that surgical treatment of those deficiencies be planned after adolescence [441].

These extreme distinctions of when to intervene point to differences in bone growth and its control in different parts of the skeleton. Treatment protocols cannot be adapted from one bone and be proper for another without regard for those considerations.

Developmental Dysplasia of the Hip.   Infants may be born with a dislocation of a hip with a frequency of 1 of 1,000 births. In milder cases, there may be a subluxation, or partial dislocation, of the joint. The cause of the joint laxity is unknown, but there may a genetic predisposition because of family history. Bracing in infancy is usually sufficient to permanently reposition the joint, but if left untreated, it may require extensive surgery. If not corrected, malalignment may result in osteoarthritis in the teen years [442].

Congenital Pseudarthrosis.   A false joint in the shaft of a long bone may manifest at birth or in infancy. Histological examination reveals that usually the gap in bone is filled with dense, fibrous connective tissue. Treatment requires removal of the fibrous tissue, fixation, and engraftment.

5.1.1.2   Cystic and Segmental Deficiencies

Bone cysts occur naturally and because they have a risk of fracture, the treatment goal is to generate bone fill with bone grafts. Other osseous voids may be a result of poor fracture healing or a result after surgical removal of diseased or malignant tissue. Many focal skeletal disorders show different prevalence in children and adults. The pediatric disorders are rare and may arise from deregulation of a growth plate.

Unicameral or solitary bone cysts are benign fluid-filled cavities in bone that are usually discovered by chance on an x-ray. They occur in children and common sites are the mandible, proximal humerus, and proximal femur [443]. They can enlarge (active cyst) or be stable (latent).

Simple lesions can be treated successfully with a series of corticosteroid injections [444]. Because the cavity can engender a risk for fracture, depending on the size and location, it may need to be treated with curettage and packing with autogenous or allogeneic bone graft or other material.

Aneurysmal bone cysts are benign osteolytic lesions with a thin wall, containing blood-filled cystic cavities separated by connective tissue septa with trabeculae or osteoid tissue and osteoclast giant cells [445]. The term aneurysmal refers to radiographic appearance. They can occur in all bone and cause extensive bone destruction. The cause is unknown, but they give the appearance of a vascular deformity. They are treated with surgical excision, rigid fixation, and reconstruction. They have a propensity to recur [446]. Data suggest that risk of recurrence is high for children with lesions in the physes or periarticular regions [447].

Giant cell tumors are benign solitary tumor that occur in the ends (epiphyses and metaphyses) of long bones or in facial bones, and contains large, multi-nucleated cells that are called giant cells. They can cause pain and tenderness in the affected area. They can expand and fracture the bone from within. They can be treated with curettage and/or cryotherapy, followed by grafting [448]. They can recur and cause deformity.

Osteoid osteomas are small, benign, bone-forming tumors that occur most often in the femur, less often in the spine. They are painful and are associated with deformities and limited mobility if they occur near a joint. They are treated by surgery or less invasive procedures such as percutaneous radiofrequency ablation, a procedure that kills the tumor cells by heat [449]. Prognosis is excellent, but they can recur and require surgery and reconstruction.

Chondromas are benign growths of cartilage that form in the small bones of the hands and feet. Echondromas are those found in the medullary canal of the bone. Periosteal chondromas are those that form on the surface of the bone. They render the bone at risk for fracture. Although they can occur as isolated lesions, they can be associated with syndromes such as Ollier's disease and Maffucci's syndrome [450]. They are treated with curettage and reconstruction. Recent work suggests benefit with radiofrequency ablation [446].

An adamantinoma is a rare, slow growing cancerous bone tumor that is most often found in the mandible or tibia. They are extremely rare, slow growing, low-grade malignant tumors, probably of epithelial origin. Symptoms include pain and tenderness and limitations in mobility. Radiographically, they appear as multicystic osteolytic lesions with a "soap bubble" character. In 20% of cases, adamantinomas can metastasize to lungs, lymph nodes, or to other bones and organs. Treatment options include limb-salvage surgery or amputation, depending on the involvement of vessels and nerves [451].

Chondrosarcoma is a cancer that occurs in adults. The lesion resembles cartilage tissue. It occurs as a primary tumor or may be secondary to a benign chondroma. There may be a genetic or chromosomal component that predisposes certain individuals to this type of malignancy. It may be difficult to distinguish low-grade chondrosarcomas from benign enchondromas by histological examination of biopsies. They may progress to high-grade, destructive tumors that metastasize and have a poor prognosis. Surgery is the main treatment option for this disease because a chondrosarcoma typically does not respond to radiation and chemotherapy as well as do soft-tissue sarcomas [452].

Osteosarcoma, or osteogenic sarcoma, is the most common bone cancer among children, adolescents, and young adults. Osteosarcoma cells can spread to the lungs and other soft tissues. Genetics may play a role in osteosarcoma, and they can occur in bones exposed to therapeutic ionizing irradiation for other types of cancer, such as Hodgkin's disease and non-Hodgkin's lymphoma. They are treated with chemotherapy, surgery and reconstruction, or amputation [453].

Ewing's sarcoma is the second most common bone cancer in children and adolescents. It may be localized to femur, tibia, or humerus or metastasize to other areas. Treatments include chemotherapy, radiation therapy, or surgery, either limb-sparing or amputation [454]. Recent investigations with autologous and allogeneic (umbilical cord) hematopoietic stem cell transplantation for tumors like Ewing's show potential and advantages especially for high-risk patients, non-responders to conventional therapy, or those who experience severe side effects from chemotherapeutics [455].

Metastatic bone disease refers to tumors that target metastatic lesions to the skeletal. The five most common carcinomas that metastasize to bone are breast, renal, lung, thyroid, and prostate. They are painful and lytic tumors can result in pathological fractures. Radiation therapy is used for palliative treatment of bone metastases. Single and multiple fraction protocols are used and re-treatments may be needed [446].

5.1.1.3 Osteoradionecrosis

Radiation therapy is used for patients with aggressive soft tissue cancers originating in the oral cavity, the oropharynx, the hypopharynx, and the larynx. In 5 to 15% of patients receiving head and neck radiation, adjacent bone undergoes gradual necrosis, termed osteoradionecrosis (ORN) [456]. ORN is a serious complication of head and neck cancer treatment. ORN is a delayed radiation-induced injury, characterized by bone tissue necrosis and failure to heal. A possible mechanism entails hypoxia, hypovascularity and hypocellularity, with vascular fibrosis and thrombosis, marrow damage, loss of osteocytes, and results in impaired bone and soft tissue healing. Other theories emphasize stem cell depletion and radiation-induced fibrosis [457].

Hyperbaric oxygen is being used at some clinical centers for osteoradionecrosis. It entails the patient breathing 100% oxygen while exposed to increased atmospheric pressure in a chamber. A series of between 30 and 40 one-hour treatments, called dives, with hyperbaric oxygen sufficient to elevate tissue oxygen tension has been proposed to prevent or to treat ORN [458]. The therapeutic principle of hyperbaric oxygen lies in its ability to dramatically increase partial pressure of oxygen in the tissues of the body. In spite of success in some series of patients, there remains a debate about the usefulness of hyperbaric oxygen for ORN [459, 460].

The devitalized, avascular bone in ORN is subject to fracture. Management of radiation-induced fractures is difficult, with prolonged healing times and a high nonunion rate. Vascularized fibula graft may be indicated for persistent nonunion. To prevent refracture, fixation should be left *in situ* indefinitely [461].

ORN is rare in patients who receive less than 60 gray (Gy) radiation therapy. The overall incidence of ORN has decreased over the last 3 decades, because of improvements in protocols and is likely to decrease further with attention to fractionation of doses and minimization of scatter [462] and use of antioxidants with radiation therapy [463].

5.1.1.4 Avascular Necrosis

Avascular necrosis [AVN], also known as osteonecrosis or aseptic necrosis, is the focal, progressive destruction of bone due to deterioration of the bone vasculature, resulting in bone and bone marrow infarction and necrosis. It is associated with the short-term use of high-dose corticosteroids, but pathophysiological mechanisms are not clear. The femoral head is the common site for AVN and may have some anatomic vulnerability. Theories include fat embolism or otherwise compromised blood flow to the marrow [464]. Patients at risk for AVN upon using corticosteroids are those with histories of hyperlipidemia, alcoholism, and tobacco abuse. Animal models have been developed with combinations of the risk factors identified in patients [465].

A growing number of reports in the literature describe the focal occurrence of osteonecrosis in the jaw, as defined by exposed, necrotic jawbones that are resistant to conventional therapy for more than 8 weeks. Analyses suggest that they are associated with bisphosphonates used to prevent bone resorption, particularly those given intravenously [466]. It has been called osteonecrosis of the jaw (ONJ), bisphosphonate-related osteonecrosis (BRON), and bisphosphonate-induced osteonecrosis of the jaw (BIOJ). Intravenous bisphosphonate therapy has significantly reduced the number of fractures secondary to osseous metastatic disease. There are now large numbers of cancer patients who have been receiving intravenous bisphosphonate therapy for more than 5 years [446]. The question arises whether bisphosphonates can over-suppress bone resorption and cause osteonecrosis by inhibiting bone remodeling. It is difficult to attribute the disorder solely to the bisphosphonate because of the frequency of other associated exposures, namely chemotherapy (55%), corticotherapy (32%), and estrogen therapy (9%). More recent evidence shows the occurrence of unusual mid-shaft fractures in patients receiving bisphosphonates [467]. Both of these complications raise concerns about the danger of excessive osteoclast suppression on remodeling and repair of bone.

5.1.2   SYSTEMIC DISORDERS OF BONE

5.1.2.1 Congenital

Osteogenesis Imperfecta (OI) describes a series of genetic disorders that result in bone fragility [468]. OI is also known as "brittle bone disease" because any bone in the skeleton can fracture with minimal stress. It can be identified as early as birth if intrauterine or birthing fractures are detected. More than 800 mutant genes have been identified in OI patients, with more than 90% of cases having dominant mutations in the genes encoding either the pro-$\alpha 1$ or pro-$\alpha 2$ chains of type I collagen, the major matrix protein in bone. Different mutations result in different severity of the disease. Other cases are associated with recessive mutations involving post-translational modifications in collagen [469].

Because it is a life-long disease, treatments focus on minimizing fractures and maximizing mobility. Based on histomorphometric studies that showed a high bone turnover rate in patients with OI, antiresorptive agents like the bisphosphonates have been tested and do show a net gain in bone mass in these patients [470].

There is a growing number of models in mice and zebrafish that resemble different types of OI [471]. It is an enormous challenge to cure such genetic diseases that involve all the bones, but research built on the growing understanding of the disorders now focuses on developing treatments that involve tissue engineering, marrow transplantation, and gene therapy to suppress mutant genes and replace with correct ones [472].

5.1.2.2 Nutritional

Osteomalacia is a histological term that refers to bone that is inadequately mineralized. If this occurs in a child, it is termed rickets and involves undermineralization of both cartilage and bone [473]. Because of reduced bone strength during skeletal growth, it manifests as bowing of the legs, fractures, and other skeletal deformities. Nutritional rickets is still encountered in infants in the United States, especially in breastfed, dark-skinned boys and girls weaned to diets rich in juices rather than dairy products [474]. In adults, the term osteomalacia is used to describe undermineralization of bone and is also associated with fractures [473]. There is bone tenderness and associated muscle weakness in these diseases. In both children and adults, the common causes are inadequate intake of calcium and/or vitamin D. Anticonvulsant drugs, e.g., phenobarbital and phenytoin accelerate the metabolism of vitamin D, and may thus cause an insufficiency. Vegan and lacto vegan's diets may not include sufficient calcium and vitamin D. Thus, nutritional rickets and osteomalacia are curable by improved diet or supplementation with calcium and vitamin D. In the United States, dairy products are supplemented with vitamin D; individuals with lactose intolerance, malabsorption, or kidney disease may need additional intake [475]. In the presence of vitamin D deficiency, the remodeling of bone results in the formation of undermineralized matrix replacing resorbed bone. It is understood that bone pain is due to the relatively increased hydration of osteoid, compared with mineralized bone, applying pressure onto the innervated periosteum. It may be difficult to diagnose osteomalacia in adults; presenting symptoms may be bone pain, muscular weakness, change in gait, or fatigue [476]. The characteristic radiographic feature of osteomalacia is the presence of pseudofractures called Looser's zone. They are incomplete fractures and can be in different states of repair. Pseudofractures can progress to a complete fracture in load-bearing regions. Bone density measurements by dual-energy x-ray absorptiometry (DXA) evaluate mineral density in a projected area and thus do not distinguish between osteomalacia and osteoporosis. Serum alkaline phosphatase may be elevated but is not a specific diagnostic marker. Diagnosis is likely if serum 25-hydroxyvitamin D is low and the ion product of calcium x phosphate is low. Vitamin D deficiency is called a silent disease, because symptoms may be diffuse, but it is a common cause or secondary osteoporosis and fragility fractures. Vitamin D deficiency was found to be more common in postmenopausal community-dwelling American women with fragility hip fracture than in control subjects [477]. Oncogenic osteomalacia is a rare syndrome of undermineralized bone, associated with tumors that secrete phosphotonin factors (FGF-23, the most common one) that cause excretion of phosphate and poor bone mineralization [478]. The osteomalacia is reversed upon removal of the tumor.

5.1.2.3 Metabolism

Hyperparathyroidism is an excess secretion of parathyroid hormone (PTH), which regulates calcium, phosphorus, and vitamin D levels, from the parathyroid glands, which are 4 small endocrine glands within thyroid tissue. It may be primary or secondary [479]. The overproduction of PTH can result in hypercalcemia and precipitation of calcium phosphate in soft tissues throughout the body. In primary hyperparathyroidism, enlargement of one or more glands is responsible for the increased production and secretion of PTH. Parathyroid carcinoma is very rare [479]. In secondary hyperparathyroidism, PTH secretion is stimulated because of low calcium and PTH acts to correct the serum calcium. Low calcium may be due to low intake or to low absorption of calcium because of vitamin D deficiency or malabsorption diseases. Kidney failure is also associated with secondary hyperparathyroidism; because diseased kidneys cannot activate vitamin D, a form of vitamin D deficiency and secondary hyperparathyroidism results. Elevated PTH stimulates bone resorption and chronic conditions can lead to osteoporosis. Although osteoporosis and osteoporosis may be generalized, there may also be focal osteolytic cysts in hyperparathyroidism.

Osteoporosis is a condition in which bones become weak and can break with very low energy, such as from falling for a standing height [480]. Osteopenia is a less severe amount of bone loss, but also carries a risk of fracture. The most common way to measure bone mass is by bone mineral density (BMD) with dual-energy x-ray absorptiometry (DXA). Values of BMD are expressed as a T-score or the standard deviation between a subject and the references average for a group of healthy young adult subjects of the same gender and race. A diagnosis of osteoporosis is made, according to the criteria of the World Health Organization, if the BMD at any site is reduced 2.5 standard deviations below the mean value for that reference group, in other words a T-score < -2.5 [481]. Osteopenia is a milder loss of bone mass, defined as a BMD value reported as a T-score between -1 to -2.5. The cellular process by which bone becomes porous with age is understood [482]. Although there can be high and low turnover variants of osteoporosis, but both involve an imbalance between bone resorption and formation. Both men and women experience age-related loss of bone from the imbalance in the remodeling process. Upon the menopause and decrease in estrogen concentrations, there is an acceleration of bone loss. In cortical bone, there is an increase in porosity, and in trabecular bone, there is a thinning and penetration of the bony struts. Loss of a bone's mechanical strength decreases its load-bearing capacity. Small changes in bone density can dramatically influence bone material properties because the elastic modulus and strength of trabecular bone are related to density by a squared relationship [483]. There are many diseases and medications known to cause secondary osteoporosis, such as hyperparathyroidism, rheumatoid arthritis, or long-term corticosteroid or phenytoin use. Current prevention/treatment modalities include anti-resorptive drugs such as the oral or annual intravenous bisphosphonates, denosumab, calcitonin, estrogens, and estrogen-like compounds with high affinity for bone and low affinity for breast and uterine tissue such as the Selective Estrogen Receptor Modulators (SERMs), like raloxifene [484]. Other classes of medications are directed at stimulating bone formation, including an osteo-anabolic form of PTH, teriparatide. For both prevention and treatment, it is necessary to ensure calcium and vitamin D sufficiency, to maintain an exercise program, and to establish fall-prevention plans.

5.1.2.4 Medications that Influence the Skeleton

The long-term use of glucocorticoids for chronic inflammatory diseases including arthritis, asthma, Crohn's disease, lupus, and other diseases of the lungs, kidneys, and liver, and for organ transplant recipients can lead to loss of bone density and fracture. Glucocorticoid-induced osteoporosis (GIO) is found in patients with pituitary tumors pituitary producing factors that increase the production of cortisol by the adrenal gland. Far more common is GIO resulting from pharmacological doses of glucocorticoids used for their anti-inflammatory and immunosuppressive effects [485]. Patients taking glucocorticoids for as little as three months may be at risk for osteoporosis [486]. Gluco-corticoids affect nearly all systems in the body and high doses invoke multiple mechanisms leading to osteoporosis. Glucocorticoids, whether taken orally or in inhaled form, inhibit gastrointestinal absorption of calcium, leading to secondary hyperparathyroidism, hypercalciuria, and hyperphos-phaturia. Glucocorticoids also directly inhibit renal reabsorption of phosphate. These dysregulated mechanisms result in increased osteoclastic bone resorption. A histopathological feature of GIO is apoptosis of osteocytes. This is also predominant in avascular necrosis associated with glucocorti-coid usage. Glucocorticoids also inhibit bone formation and osteoblast differentiation. An indirect mechanism is by inhibition of osteo-anabolic sex hormones, estrogen, testosterone, and adrenal de-hydroepiandrosterone. There are direct effects of glucocorticoids on osteoblast determined *in vitro*: high doses of glucocorticoids inhibit synthesis of bone matrix components, although in many model systems, low doses stimulated matrix synthesis [487].

Thyroid hormones, when taken in excess or when produced in excess, are associated with osteoporosis. Thyroid hormones are essential for normal skeletal growth and the maintenance of bone mass in adulthood. Experimental studies indicate that thyroid hormone promotes chondrocyte differentiation, chondrocyte maturation in the growth plate, osteoblast activity both directly and indirectly, and osteoclast differentiation by increasing osteoclastogenic factors in osteoblasts [488]. In children, hypothyroidism causes impaired bone formation and growth retardation whereas thy-rotoxicosis results in accelerated growth, advanced bone age, and decreased bone mass.

Other drugs that are associated with secondary osteoporosis are certain antiseizure drugs, such as phenytoin and barbiturates, gonadotropin-releasing hormone (GnRH) drugs used to treat endometriosis, excessive use of aluminum-containing antacids, heparin, lithium, methotrexate, and selective serotonin reuptake inhibitors (SSRIs) [489].

5.1.2.5 Co-Morbidities that Influence the Skeleton

Diabetes is associated with a number of bone problems [490]. There are two types of diabetes. Type 1 diabetes is an autoimmune disorder, in which the destruction of the pancreatic beta cells results in an absence of insulin. Type 1 diabetes occurs most commonly in juveniles but can occur in adults. In type 2 diabetes, which is far more common, there is hyperglycemia because of resistance to the actions of insulin in muscle, fat, and the liver and an inadequate response by the pancreatic beta cell. There are increased risks for low bone mineral density and for bone fractures in patients with type 1 and, to a lesser extent, type 2 diabetes mellitus [491]. These abnormalities may be related, at least in part, to prolonged periods of poor metabolic control. The mechanisms for low bone density are complex. In type 1 diabetes, absence of insulin and insulin-like growth factors (IGFs) during growth may be the cause of low bone formation. These patients may also have malabsorption disorders, like celiac disease, which is also associated with osteopenia and osteoporosis. Evidence indicates that there is low bone turnover in both types, with improvement accompanying metabolic control. Clinical studies have also demonstrated a significantly higher incidence of delayed union, nonunion, and a doubling of the time to healing of a fracture in diabetic compared with non-diabetic patients [492]. Animal models of diabetes also show delayed fracture healing [493].

There are several aspects of diabetes that contribute to diabetic bone disease [494]. Low circulating IGFs, high cortisol, and possibly low PTH contribute to low bone mass and poor bone formation. Poor blood flow is a characteristic of diabetes and was detected in bones of diabetic rats [495]. Nonenzymatic glycosylation and oxidation of proteins in serum and in tissues occurs with aging and is accelerated in diabetes. They are called advanced glycation end products (AGEs) and their receptors are called RAGEs. They inhibit osteoblast function and RAGEs were upregulated and associated with delayed bone healing in diabetic mice [496].

There is accumulating evidence that women, but probably not men, who take one of the thiazolidinediones, commonly used to treat insulin resistance in patients with type 2 diabetes, have a greater risk of developing weakened bones and fractures [497]. It is unfortunate that the very drug that promotes insulin sensitivity in fat cells also inhibits osteoblasts and stimulates osteoclasts, two effects that result in increased bone loss [498]. The class of antidiabetic drugs, thiazolidinediones, causes bone loss and further increases facture risk in already at-risk diabetic patients; thus, the thiazolidinediones are in the category of drugs causing secondary osteoporosis [499]. In addition to the association with increased fracture risk, thiazolidinediones also delayed bone healing in experimental animals [499].

Renal osteodystrophy refers to all the bone and mineral disorders that occur in patients with chronic kidney disease (CKD). There is a wide spectrum of skeletal manifestations that ranges from those associated with low turnover (osteomalacia, adynamic bone) to those associated with high turnover (bone loss) [500]. In low-turnover renal bone disease, hypocalcemia and osteomalacia were once common because of reduced renal synthesis of 1,25-dihydroxyvitamin D; they are less common now because of the use of analogs of vitamin D to restore mineral homeostasis and bone matrix mineralization. High turnover renal bone disease is associated with hyperparathyroidism and its skeletal manifestations. Bone fractures are common in patients receiving dialysis treatments for advanced renal disease [501]. Many patients with CKD also have diabetes, and they are at greater risk for low turnover bone disease; it has been suggestion that vascular disease may be at the root of both CKD and diabetes [502].

5.2   CARTILAGE

5.2.1   OSTEOARTHRITIS

5.2.1.1   Clinical Features

More than 1 million joint replacement procedures are performed annually in the United States for hip, knee, and shoulder deformities, with more than 90% due to degenerative osteoarthritis (OA) [395]. There are different variants of the degenerative osteoarthritides, resulting from aging processes, congenital anatomical abnormalities, trauma, obesity, and secondary to growth hormone deficiencies and diabetes, but all are associated with erosion of the cartilage from the surface of a joint. Recent analyses suggest that there are three major subsets of OA based upon pathophysiological mechanisms [503]. Type I OA is the rapidly progressing type with various genetic predispositions; Type II OA is that which is associated with the menopause, and Type III is the prototypic aging-related OA. Genetic research with families presenting with rapidly advancing OA before the age of 40 reveals several groups of mutations, involving structural matrix molecules or regulatory genes or determinants of skeletal shapes or polymorphisms in genes that regulate chondrocyte differentiation or function. The high female-to-male ratio of 10:1 for knee OA, with a peak for women either perimenopausally or within 5 years of natural menopause or hysterectomy support the importance of estrogens in OA [504]. Although there may be many different pathophysiological mechanisms and potential therapeutic approaches, all osteoarthritides present with degenerative joint pain and disability.

OA is a painful degenerative joint condition in which there is a loss of articular cartilage and compensatory osteophytes (bony outgrowths, spurs) and subchondral osteosclerosis (a thick layer of dense bone) [5]. OA is the most common degenerative joint disorder in humans and most other mammals. Although it is associated with aging, many older individuals have perfectly healthy and functional articular cartilages. There are also many elderly individuals who are asymptomatic but have radiographic evidence of joint space narrowing and osteophytes [505]. The incidence of OA in patients over 45 years of age is dramatically increased, with approximately 60% of the population over the age of 60 years affected. OA involves the gradual destruction of joint cartilage, and it mainly affects weight-bearing joints such as the knees, hips, and spine. It can also affect the hand and shoulder. OA progresses gradually and is usually monoarticular. Its causes are unknown and its pathogenesis is not well understood. Besides aging, factors such as genetics and altered joint mechanics due to anatomy or to previous injuries increase the probability of OA occurrence. Patients with OA experience pain, swelling, deformity, joint enlargement, and limitation of joint motion, among other symptoms. The first lines of nonpharmacological therapies, such as exercise, weight control, and over-the-counter supplements of glucosamine sulfate and chondroitin sulfate, can make many patients more comfortable. The symptomatic treatment of OA includes non-steroidal anti-inflammatory agents and injections of hyaluronan or corticosteroids. These may address the symptoms for some patients but do not halt the progression of the disease. Eventually, surgical intervention in the form of total joint replacement may become necessary to overcome pain and dysfunction.

Characteristic degenerative features of OA articular cartilage include the development of fissures and focal erosive lesions that are diagnosed by arthroscopic examination [506]. Radiologically, there is narrowing of the joint space, increased subchondral bone thickness (osteosclerosis), and formation of periarticular cysts and large osteophytes. Studies with animal models show that the earliest sign is loss of proteoglycans from the articular surface. The initial stage may progress to the fibrillation stage where cracks develop perpendicular to the articulating surface. Pieces of cartilage can dislodge into the synovial space and cause inflammation and foreign body giant cell reaction. The cracks can extend from the cartilage surface to the underlying bone where they elicit neovascularization and subchondral bone osteoclastic resorption. The subchondral bone plate thickens around the area of the crack. Neovascularization brings in mesenchymal cells and induces fibrocartilage formation; however, the fibrocartilaginous repair tissue is of poor quality, and as the disease progresses, it may not persist. Exposure of the subchondral bone results in rasping against the opposing articulating surface, and eventual eburnation [507]. Complete erosion of the articular cartilage is associated with dense sclerosis at the bony surface, with fibrous and fibrocartilage tufts. There is new bone formation, but also areas of osteonecrosis. Osteophytes covered with normal-appearing hyaline cartilage form at the bony margins of the joint. They alter the joint contour and may limit mobility. Cysts also develop, most often close to the joint surface. The cysts contain a gelatinous material similar to that found in ganglions adjacent to tendon sheaths [3]. Radin and Rose advanced an alternate idea about OA pathophysiology that damage in the subchondral bone occurs first, possibly due to overloading or misalignment [508]. These interstitial microfractures heal with callous formation and remodeling that alters the biomechanical environment in the overlying cartilage and results in horizontal splitting and deep fibrillation of the cartilage.

There is a growing body of evidence for genetic predispositions to OA involving an array of genes [509]. The clustering of OA in families, in sibling-pairs, and in twins has led to genetic association studies that indicate predisposition to OA in individuals with polymorphisms in collagen II, in genes known to control skeletal development, such as the bone morphogenic proteins (BMPs) and WNTs, genes concerned with apoptosis, inflammation, and neurotransmitters involved in pain perceptions. Genome-wide association studies (GWAS) are intended to identify find variants in specific genes or narrow genomic regions that are associated with the presence or severity of a specific clinical condition. They require large numbers of genomic DNA samples linked to definitive diagnoses, but they do not presuppose a biological mechanism underlying a genetic association. Such a study with a pool of international subjects led to the identification of polymorphisms in a gene called PTGS2, which encodes for COX-2, a key enzyme that catalyzes the synthesis of prostaglandin E2 [510].

5.2.1.2 Common Pathophysiological Processes

Chondrocytes in immature individuals have the capacity to divide and enlarge the cartilaginous structure. There is evidence for synthesis of DNA and of collagen, proteoglycans and other ECM molecules [346]. With cessation of limb growth, normal cartilage chondrocytes lose the capacity to divide. The modest capacity of the articular chondrocyte to repair and remodel the ECM diminishes further with age [511]. Studies with OA cartilage show normal or slightly increased cellularity that depends on the stage of disease [512]. Other studies show chondrocyte apoptosis in OA [513], with apoptotic cells primarily in the surface and middle zones and closely associated with structural changes in the matrix and extent of degradation [514].

In response to injury, there is limited proliferation of chondrocytes that are seen in histological specimens as clusters or nests of cells [515]. Initially, OA chondrocytes may be more metabolically active than those from normal articular cartilage [516]. OA chondrocytes incorporate proteoglycans and synthesize hyaluronate at rates that are higher than those in normal tissue, but there are alterations in the ability of the secreted proteoglycans to interact with the hyaluronate or to aggregate [517]. The synthetic response is a weak reaction, relative to the degradation of the synthesized product. Collagen and proteoglycan synthesis are increased in OA cartilage, but so is proteoglycan breakdown, possibly due to increased activity of the enzymes that are already present and active in normal tissue [518]. Before its complete erosion, articular cartilage can be graded into zones of disease severity by gross examination (Outerbridge classification [519]) and by histological features (Mankin classification [520]). With surgical samples thus classified, there was a striking decrease in gene expression of aggrecan and collagen II that was correlated with increase in disease severity [234].

Although there is evidence of increased anabolic activity in early stage OA, the overwhelming feature of arthritis is the conversion of chondrocytes to catabolic cells. Eventually, the degradative process is greater than the repair process, resulting in the end-stage disease and total loss of cartilage [46]. Especially in animal models of OA, synovial fluid, blood, and urine contain markers of collagen synthesis and degradation and of proteoglycan synthesis and degradation [521]. Analyses of human OA cartilage have shown increased amounts of a number of proteins indicative of the catabolic state, including matrix metalloproteinase and aggrecanases [522, 523]. In addition, OA cartilage features increases in other degradative enzymes such as cathepsins [524] and stromelysin [525], whereas the levels of tissue inhibitor of matrix metalloproteinases are not increased [523]. These disparities support the hypothesis of a disruption in tissue homeostasis in OA cartilage. Further, the inflammatory mediator interleukin-1 is elevated in OA cartilage where it stimulates enzyme synthesis and catabolic activity [525].

The changes in the composition and ultrastructural organization of articular cartilage during OA result in changes in the material properties. Osteoarthritic cartilage shows increased swelling upon soaking in Ringer's solution, and it is both weaker in tension and softer in compression when compared with healthy cartilage [285].

5.2.1.3 Relationship Between Osteoarthritis and Osteoporosis

Both OA and OP are diseases that increase the age, and it had been taught that they are mutually exclusive. That notion was reasonable in light of OA being associated with increased body weight and OP being associated with low body weight. In 1972, Foss and Byers noted the radiographic absence of osteoarthritic changes in fractured hips of osteoporotic subjects and stated "that osteoporosis and osteoarthritis of the hip do not normally occur together" [526]. Osteophytes and subchondral osteosclerosis can give the impression of increased bone density at the hip, as measured by dual-energy x-ray absorptiometry; that addresses the relationship between OA and nearby bone, but does not deal with the question of systemic OP [527]. In a study that was designed to compare osteoporotic women with a control group of women with OA, we were surprised to find that 25% of the 68 women with advanced OA had occult OP, defined as a T-score at any site of less than −2.5, and that they had elevated markers of bone turnover, including urinary N-telopeptides and serum bone-specific alkaline phosphatase and osteocalcin [528]. Other subsequent studies made similar observations (reviewed in [482]). Although the typical OA presents with greater body weight, lower body mass index, bone density, and focal osteosclerosis, a subset of OA patients has low bone density at skeletal sites not involved with OA. The OA subjects with occult OP had lower body weight, body mass index, and advanced age, compared with OA patients without OP [529]. A more recent analysis of bilateral hip bone density scores in subjects with unilateral hip OA revealed that, although the OA hip may have elevated bone density, in 32% of the subjects the contra-lateral (non-OA) hip was classified as osteopenic (T-score between −1.0 and −2.5) [530]. That discordance between hips may have a negative impact on patient treatment decisions if clinicians are unaware of the possibility of OP in patients with OA [531].

5.2.2 RHEUMATOID ARTHRITIS

Rheumatoid arthritis (RA) is a chronic, systemic autoimmune disease that manifests itself in the joints. RA affects the joints in a symmetric pattern with inflammation, pain, stiffness, joint destruction, deformity, and disability. Systemic symptoms include fatigue, malaise, and morning stiffness. Joint symptoms occur in periods of flares, a phenomenon that complicates investigations. Structures other than the joints such as skin, heart, lungs, and eyes may also be affected, with the cardiovascular system bearing the most burden [532]. RA affects three times more women than men, and it generally presents between the ages of 40 and 60 [533], although it may manifest itself at any age. When it occurs in children younger than 16 years of age, it is identified as a form of juvenile RA [534]. The prevalence rate of RA is 0.5 to 1% worldwide. It affects all populations, but it is more prevalent in groups such as Native Americans and less prevalent in Caribbean blacks.

The cause of RA is not known, but a significant genetic component is widely suspected. There are genetic and non-genetic components, as suggested by a 2-to-3 fold increased risk of RA in first-degree relatives of RA patients, and by 4.3-fold in monozygotic twins [532]. As many as 90% of RA patients have the epitope of the HLA-DR4/DR1 cluster, compared with 40% of controls. Immunological testing shows abnormal antibodies in the serum of RA patients; 50 to 80% of patients have elevated serum levels of Rheumatoid Factor (RF), a non-specific autoantibody, or of anti-cyclic citrullinated peptide (CCP) antibody [535]. During inflammation, arginine amino acids in tissue proteins can become converted to citrulline; susceptible individuals develop auto-antibodies to these altered proteins. A case-control study of a series of subjects who had donated blood to a blood bank revealed that antibodies to CCP and RF were present in 34% and 17%, respectively, of those who later developed symptoms of RA [536]. A finding of elevated serum level of anti-CCP or anti-RF in a healthy individual implies a high risk for the development of RA. It is clear that dysregulation of B-cell/T-cell interactions drives the inflammatory process through local upregulation of tumor necrosis factor (TNF). Emerging epidemiological evidence indicates a strong association between vitamin D-deficiency and autoimmune diseases, including RA [537]. Experimental data show that the vitamin D receptor (VDR) is induced in activated T-cell, B-cells, and macrophages, and that 1,25-dihydroxyvitamin D and analogs inhibit proliferation and production of cytokines by those cells [538]. Small trials indicate the clinical benefit of achieving vitamin D-sufficiency in patients with RA [539] and larger intervention studies are underway.

Smoking is the dominant environmental risk factor in RA (reviewed in [535]). Hormonal components are also thought to be involved because the disease increases with age and is more frequent in women. Infections with virus or bacteria have also been suggested as causes of RA, although a particular offending organism has not been identified [540]. The proposition is that, in genetically predisposed people, certain viral infection can trigger a distorted process by which patients produce antibodies that erroneously identify and attack their own tissues, a concept called molecular mimicry [541].

RA affects the small joints of the hands and feet in a symmetric distribution. The most commonly affected joints are those of the hand and wrist followed by the knee, shoulder, ankle, cervical spine, hip, elbow, and temporomandibular joints. The affected joints become swollen, warm, red, and tender with decreased range of motion. Destruction of the joint and tendon structures leads to the characteristic deformity of RA joints. Other musculoskeletal consequences of RA include tenosynovitis and associated tendon rupture, periarticular osteoporosis, and carpal tunnel syndrome [532].

The pathology of RA begins with synovial cell hyperplasia and endothelial cell activation. These events progress to uncontrolled inflammation and vast proliferation of synovium, which result in the destruction of local cartilage, bone, tendons, ligaments, and blood vessels. Many cells are involved in the development of RA, including CD4 T cells, B lymphocytes, mononuclear phagocytes, fibroblasts, neutrophils, and osteoclasts. The B-cells are responsible for the production of autoantibodies. RA is associated with abnormal production of inflammatory molecules [532]. Overproduction and overexpression of tumor necrosis factor alpha and IL-6, in particular, drive synovial inflammation and joint destruction [542, 543].

There are many diagnostic classifications for RA. The common criteria are morning stiffness, involvement of 3 or more joint areas, swelling of hand joints, symmetric joint involvement, rheumatoid nodules, serum rheumatoid factors, and radiographic changes on posteroanterior hand and wrist radiographs. The clinical symptoms must be present for at least 6 weeks [532].

The diagnosis of RA is associated with significant morbidity, disability, and mortality. Individuals with RA become gradually impaired in their ability to perform activities of daily life. Five years after the diagnosis is made, as many as one third of RA patients are unable to work. RA shortens life expectancy by 5-10 years. Those with poor functional status, advanced age, male sex, positive rheumatoid factor findings, extra-articular disease, elevated acute-phase response, increased involvement of joints, infections, cardiovascular disease, renal disease, GI bleeding, and lymphoproliferative disorders are at increased risk for mortality [532]. Patients with RA have a 60% greater risk of dying of cardiovascular disease than the general population [544].

Although RA cannot be cured, novel therapies help mitigate the destructive effects of the inflammatory processes. Major improvements in the outcomes of RA began with the use of methotrexate in the 1980s [545], and shortly after, it was found that this disease-modifying antirheumatic drug (DMARD) could effectively be used in combination with other agents that have different mechanisms of action [546]. Since then, there has been introduction of an increasing array of biological agents such as anti-tumor necrosis factor alpha (TNFα) monoclonal antibody [547]. Currently, DMARDs are used in combination with analgesics, cyclooxygenase-2 inhibitors, and non-steroidal anti-inflammatory drugs (reviewed in [548]). Methotrexate continues to be the DMARD of choice [535]. Biological agents currently used include TNFα-inhibitors, and monoclonal antibodies against CD20, IL-6, and CTLA4-Ig. However, side effects and the high costs of biological agents preclude their widespread use [535].

5.2.3   OSTEOCHONDRITIS DISSECANS

Osteochondritis Dissecans (OCD) is a term that describes an osteochondral fracture. A piece of cartilage with a thin layer of subchondral bone breaks from the end of a bone, often at the articular epiphysis [549, 550]. If the fragment stays close to where it detached, there may be few or no symptoms of OCD, and the fracture can often heal itself. There may be radiological or arthroscopic evidence of this. If the fragment dislocates between the moving parts of the joint, surgical repair may be necessary. OCD is caused by primary trauma and secondary ischemia. OCD most commonly affects the knee joint, but it can also occur in the elbow, shoulder, hip, and ankle joints. The symptoms are pain, popping or locking of the affected joint, joint weakness, decreased range of motion, swelling and tenderness. OCD is three times more common in males than females, and more often affects young patients aged 10 to 20 years. Sports participation increases the risk of OCD. OCD increases the risk of later OA of the affected joint [550].

In the acute injury stage, the intra- and peri-articular soft tissues become thickened and edematous. The adjacent metaphysis often shows mild osteoporosis. Later, the contour of the epiphysis may show fragmentation and become irregular with thinning of the subcortical zone of rarefaction. Because of vascular injury or microfactures of the trabeculae, the blood vessels within the epiphysis are incompetent and healing is poor. Necrotic bone cannot properly support the joint, and compression flattens the articular surface. Granulation tissue gradually replaces the necrotic tissue. The injured articular cartilage allows for synovial fluid to enter into the epiphysis, creating a subchondral cyst, which together with increased joint pressure may prevent healing [549].

Repair of symptomatic OCD lesions in the femoral condyle is usually surgical. If the lesions are stable and nondisplaced, drilling is used to induce bone healing. Large unstable or displaced defects must be fixated with compressive screws. Metal and resorbable devices are used, with metal implants removed prior to full weight bearing. Often removal of the fragment is sufficient, especially when the lesions are small. Outcomes with larger fragments are better for repair than for removal alone. Osteochondral allograft transplantation and autologous chondrocyte implantation are other therapeutic options associated with positive outcomes. Cartilage repair is more difficult for lesions in the patellofemoral compartment. Microfracture, osteochondral autograft transfer, and osteochondral allograft transplantation are more limited in the patellofemoral compartment, compared with the femoral condyle. Autologous chondrocyte implantation was initially largely unsuccessful as well in this anatomical area. Lately, owing to our improved understanding of patellofemoral joint mechanics and improved treatment of tracking abnormalities of the patella, the results of autologous chondrocyte implantation for OCD of the patellofemoral compartment have improved significantly (reviewed in [551]).

5.2.4   TEMPOROMANDIBULAR JOINT (TMJ) DISORDERS

This term encompasses a number of disorders that inflict tenderness, pain, and dysfunction in the region of the temporomandibular joint (TMJ) (reviewed in [552]). TMJ disorders are most common in women aged 30 to 50 years. TMJ disorders may involve derangements within the masticatory muscles (extracapsular, external, myofascial) or within the joint (intracapsular, internal) [553]. Myofascial pain dysfunction is the most common disorder associated with TMJ syndromes. It may be due to muscle hyperactivity from nocturnal grinding or clenching, termed parafunction. Degenerative changes in the joint occur in response to repetitive overloading. No clear relationship has been established between occlusal alterations and TMJ disease. Often, the symptoms of TMJ disorders can be treated satisfactorily with nonsurgical treatments and self-care.

In TMJ osteoarthrosis, there is deterioration and abrasion of the articular cartilage with thickening of the underlying bone [554]. The symptoms include pain and tenderness of the jaw, aching pain in and around the ear and in the face, difficulty or discomfort while chewing, locking, popping or clicking of the jaw, and referred pain and headache. A small percent of individuals with TMJ disorders call for dental or surgical intervention. It also occurs at a higher incidence in individuals with rheumatoid arthritis, fibromyalgia, chronic fatigue syndrome, or sleep disorders [554].

Abnormal disc/condyle relationships can be identified by magnetic resonance imaging, but are also common in asymptomatic individuals [555].

Most therapies for TMJ disorders are aimed for pain management. Intraoral splints and appliances are mildly effective at reducing pain [556], as are behavioral therapy, jaw exercises, and postural training [557]. Pharmacological agents having shown effectiveness for pain management include anti-epileptic drugs, nonsteroidal anti-inflammatory drugs, corticosteroids, analgesics, clonazepam, diazepam, antidepressants, skeletal muscle relaxants, capsaicin and intraarticular injections of hyalunorate, corticosteroids, opioids, and intramuscular lidocaine or botulinum toxin (reviewed in [557, 558]). Tricyclic antidepressants are effective at decreasing bruxism [559]. Hyaluronic acid injections have been recently introduced for management of pain and discomfort, with outcomes comparable to those achieved with corticosteroid injections or oral appliances [560]. Acupuncture is also sometimes useful to provide short-term analgesia [561].

Surgical options for patients with severely dysfunctional TMJ are few. There are no safe, effective TMJ disc implants; therefore, many patients undergo removal of the injured TMJ disc, which leaves the TMJ without an important functional component. Tissue engineering efforts have recently focused on producing safe, viable implants for these cases [562, 563]. Arthrocentesis, arthroscopy, arthroplasty, distraction osteogenesis, osteotomy, and joint reconstruction or replacement are surgical interventions that aim at restoring some degree of anatomical normality and function of the most severely damaged TMJ joints. Results, however, remain suboptimal [552].

CHAPTER 6

Animal Models

6.1 BONE

6.1.1 BIOCOMPATIBILITY

Animals have been used routinely to assess biocompatibility or toxicity of materials [564]. There are many ways to test for biological compatibility, but the major considerations ought to be related to the intended clinical use: the nature of the tissue to be contacted and duration of contact. Consensus standards for testing protocols are developed by organizations such as ASTM (American Society for Testing and Materials), USP (United States Pharmacopoeia), or ISO (International Standards Organization). Soft tissue sites in normal animals can reveal biocompatibility or whether there are acute or chronic inflammatory reactions to implanted materials [564].

The advantages of rats are availability and low cost, but there may be difficulties in extrapolating results to larger animals. Inevitably, there is an inflammatory response to the surgical trauma from the skin incision and from the placement of the material or device. If that response is negligible or temporary, the material is considered to be biocompatible. It is good procedure to implant the test material at a distance from the incision so that repair of the skin wound does not interfere with tissue reaction to the test material [394]. This can be done by preparing subcutaneous pockets by blunt dissection over the ventral thoracic area through an incision made at the level of the diaphragm (Figure 6.1A). The test material is then deposited in the cephalad end of the pocket and the incision closed with inert 9-mm stainless steel wound clips at least 1 cm away from the implant. Use of suture material on top of the implant would evoke its own reaction that can confound the response to test material. Histological examination of the implantation site and surrounding tissues should be done at intervals to assess tissue responses and fate of the implant. Inert (biocompatible) materials may engender nothing or a benign fibrous tissue reaction, and other materials may elicit an infiltrate of macrophages and polymorphonuclear leukocytes by day 3 or 4, and foreign body giant cells as early as 7 days after implantation in rats [394]. At approximately three days after implantation of a biocompatible material, fibroblasts produce collagen fibers that are histologically identifiable. Because factors such as the composition, size, shape, and surface character of materials can influence the nature of the foreign-body reaction, it is important that those parameters be carefully controlled when comparing materials [565]. Evidence indicates that inflammatory reactions may be related to the affinity that monocytes have for certain materials, as have been shown for polymethylmethacrylate [566]. If a component of an implant is bioabsorbable, it is necessary to evaluate its *in vivo* rate of degradation and the effects, if any, of degradation products [567]. Toxic materials are those shown to release substances that causes necrosis of the adjacent tissues.

6.1.2   OSTEOCOMPATIBILITY

Materials or devices that are biocompatible in soft tissues are not necessarily inert in bone-healing sites. Because we had found that several resorbable polymers prevented ectopic bone formation [566], it was necessary to develop a simple quantitative model of osseous bone formation to assess effects of implant materials on bone remodeling processes that are needed for implant incorporation. The goal was to develop a model that would reveal whether a material was compatible with normal bone healing. The rodent lateral tibia has a flat surface in which a unicortical wound can be made reproducibly with a low-speed, hand-held drill, without risk of fracture because only one cortex is wounded (Figure 6.1B). It shows exuberant healing within weeks after wounding. In such a healing wound, there is a reproducible sequence of intramedullary bone formation, repopulation of the medullary canal with marrow, and complete cortical regeneration by 2 weeks. We recently adapted the rat model [393, 570] to mice [571]. Materials that do not interfere with that spontaneous bone regeneration would be considered osteocompatible. If a material activates marrow inflammation or its degradation products interfere with bone wound healing in such a model, there is concern that a device made with it would not be incorporated into bone. Implantation in soft tissue sites reveals general tissue biocompatibility, whereas implantation in a milieu of active bone formation provides information about performance within regenerating bone. The small tibial wound is a model for fracture healing, in contrast to larger osseous defects that are used to test bone grafting materials and implant materials. The term "wound" describes a healing injury and the term "defect" describes a non-healing injury. The latter must be larger and documented in each study to not heal spontaneously (Section 6.1.4 below).

Another model to test for osteocompatibility is the use of bone conduction chambers, which monitor the predictable ingrowth of new bone and which showed, for example, inhibition by cement particles [568] and osteocompatibility with hydroxyapatite particles [569]. Fracture healing models have become more useful for systemic manipulations than for tissue engineering studies [570].

Another way to test for osteocompatibility is to use a subcutaneous site for implantation of an admixture of the test material with an osteoinductive material, like demineralized bone. From such testing, a number of materials were found to inhibit osteogenesis that is normally induced by demineralized bone, including polymethylmethacrylate, polylactic acid, and absorbable tricalcium phosphate [566]. Those findings suggest that an excessive inflammatory reaction by a component of an implant can undermine an intended osteogenic effect.

6.1.3   BONE SUBSTITUTES

Osteoaugmentation refers to the use of implants, devices, or constructs to increase the mass of bone in a deficient site. Experimental animal models have been developed in many species for testing bone-substitute implants or tissue engineered constructs for potential clinical applications. For potential use in orthopedic, craniomaxillofacial, and plastic reconstructive surgery, small animals such as mice and rats are often used for screening purposes and for proof-of-principle studies (Figure 6.1). Larger preclinical animal models are used to evaluate implants and devices for augmenting bone deficiencies

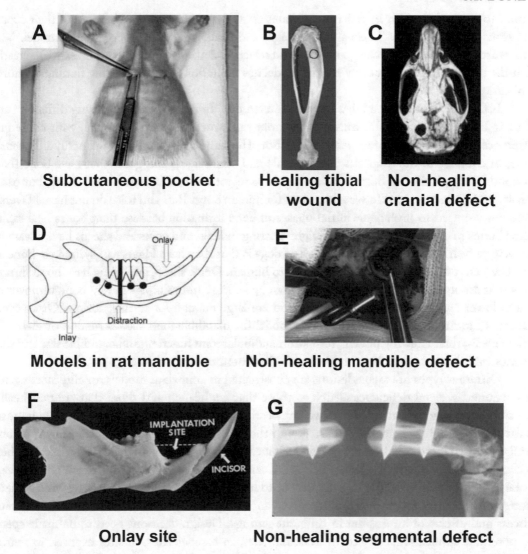

Figure 6.1: Rat models for testing devices and materials for skeletal reconstruction. (A) Subcutaneous pocket for material testing. (B) Location of healing tibial wound. (C) Dried rat skull shows non-healing defect in left parietal bone and repair of right parietal defect after implantation with an osteoinductive material. (D) Drawing shows three uses of rat mandible: a non-healing defect for inlay testing, site for onlay testing in edentulous area, and location of osteotomy and locations of fixation screws for distraction osteogenesis. (D) Intraoperative photograph shows location of mandibular defect and use of dental amalgam gun for transfer of test material. (E) Dried rat mandible shows location for onlay test material to augment the bony contour. (F) Radiograph of non-healing segmental defect in midshaft.

similar to those that occur in patients. Primates are very rarely considered because of their expense and restricted availability. There are advantages and disadvantages to every animal model. Many factors are important in choosing a species, and selection depends upon the major research question, with the understanding that every animal model has limitations in extrapolating findings to human applications [575, 576].

It should be understood that bone architecture in different species or during different stages of their life-cycles may differ considerably from the structure of human bone, with consequent differences in mechanical parameters [571]. Normal human bone is organized primarily as Haversian bone, in contrast to other species, like juvenile and adult sheep and goats that have bone that is predominantly plexiform. Some species may have regions with Haversian bone tissue, but osteon density and diameter are of lower values than for human bone. Rats and mice do not have Haversian bone but are used to investigate initial phases of bone formation because their bones heal rapidly. Some bones in cats are used for non-weight-bearing models, and goats and sheeps have advantages for weight-bearing models. An advantage of dogs is that they have Haversian systems in bone and a mechanism of internal remodeling similar to human. Dogs, sheep, and goats have bone turnover rates that are much higher than that in human bone. Pigs and minipigs have a bone turnover rate that is lower and most similar to human, and are large enough for testing devices. Other special features of swine make them an attractive model for mandible joint reconstruction or distraction osteogenesis; their chewing pattern, temporomandibular joint function, and mandible size and shape are more similar to human than are those of other common research species [372].

Different types of osseous lesions are constructed in animals as models for clinical congenital and acquired skeletal deficiencies, such as those that are healing and those that are non-healing, those that are weight-bearing or non-weight-bearing, those that are load-bearing (through muscle, as for the mandible) or not-load bearing, and those that are in segmental, intraosseous, or onlay configurations (Figure 6.1). An animal model is meant to have a correspondence to a clinical problem, a concept called connectivity. Experimentation can be designed to test performance of a device, to reveal design principles to improve devices, or to help elucidate the mechanisms by which a device successfully achieves its purpose. Standardized animal models are needed to characterize the repair process and effects of an implant in different settings. Design decisions need to be made specific for potential applications. Effects in avascular sites, in osseous sites, in sites exposed to marrow, or in weight-bearing sites, for example, may differ dramatically and may eliminate possible clinical applications. Algorithms can be developed to design and test implants and to ensure that the findings in animal studies have connectivity to a real clinical need.

In the simplest case of bone-substitute materials, a study may seek to determine whether a test material can function as well as, or superior to an autogenous bone graft. Given the different mechanisms of bone formation (osteogenesis, osteoconduction, osteoinduction, Table 4.1), it may be desirable to determine the mode by which the use of an implant or device can result in increased bone formation. Theoretically, a device with cells or engineered tissue may work by osteogenesis, in other words, by direct formation of new bone by the transplanted cells. A device may work by osteoconduction, by providing a scaffold onto which adjacent osteoblast or progenitors may migrate from the margins of a defect and deposit new bone. A device may work by osteoinduction, by providing a morphogenic signal to convert non-skeletal cells into bone cells. A device may work by osteoaugmentation, by providing factors that stimulate proliferation or activity of resident osteoblasts. Frequently, more than one of the mechanisms may be in effect with a particular implant. Testing materials in different sites can be useful to reveal tissue-specific responses, and the use of multiple sites allows for fuller characterization of bone-substitute materials [393].

To be a rigorous test for osteogenesis, or the ability of the transplanted cells or *in vitro* engineered tissue constructs to directly form bone *in vivo*, the first model should employ a non-osseous site without possible contribution of adjacent sources of bone. Good vasculature in the site is a prerequisite to ensure viability of transplanted bone cells or tissue. Subcutaneous or intramuscular implantation sites can be used to track the participation of the transplanted cells in ectopic bone formation. Inexpensive rodents are useful for time course studies, for comparison of carriers and cell density, and for investigating the fate of donor cells by cell survival, differentiation, and participation in osteogenesis and/or angiogenesis. Formerly, the fate of donor cells was tracked by exploiting interspecies differences in patterns of histological staining [572]. With current technology, donor cells can be pre-labeled by retroviral transfection prior to implantation. Labeling agents must remain absent from host cells, undiluted by cell division, easily detectable, and they must not interfere with cellular activities [573]. Examples of cell fate-tracking agents are β-galactosidase, BrdU, and enhanced green fluorescent protein (EGFP). Detection of β-galactosidase requires a staining protocol, however, and BrdU cannot ensure long-term labeling. EGFP is stable and innocuous, being detectable with ultraviolet light and fluorescence microscopy. The *in vivo* fate of donor cells has been monitored over weeks in mice models of subcutaneous implantation with EGFP-labeled osteoblast progenitors [573], with GPS-luciferase-labeled goat osteoblast progenitors [574] or in a rat calvariae model using β-galactosyltransferase-labeled murine MSCs [575]. Rarely have studies quantified specific cell behaviors such as survival, density, or contribution to osteogenesis, but use of *in vivo* bioluminescence imaging is increasing to monitor the fate of donor cells for bone engineering [574, 576, 577]. A study that transplanted live mice bone grafts expressing β-galactosidase into wild type mice showed that 70% of osteogenesis was attributed to donor cell participation in endochondral bone formation, and that remodeling of new bone and cartilage led to replacement of donor tissues/cells with host cells after 4 weeks [578]. Another study quantified the survival and density of EGFP-expressing rabbit progenitor cells seeded in gelatin sponges and implanted subcutaneously, and it reported a significant decrease in the density of donor cells over 5 weeks [579]. There was, therefore, a strong suggestion that for approximately 4 weeks following implantation, donor cells may actively participate in *in vivo* osteogenesis, after which the host cells and associated vasculature migrate into the implants and take over remodeling the newly formed bone. Cell fate and function were shown to depend upon the site and the species [580]. In mice, labeled cells seeded in porous ceramic cubes were present for 6 weeks after implantation in dorsal subcutaneous pockets, but in rats, they were detected for only 2 week. Bone was produced in the mouse model but not in the rat. Rats were also used for a spinal implant, but donor cells were detected for only 1 week, with minimal formation of new bone in the scaffold.

Subcutaneous or intramuscular sites have also been used to assess osteoinductivity of cell-free materials, such as transitional epithelia [581], demineralized bone [404], demineralized dentin [581], and bone morphogenic composites [582].

Non-healing osseous defects have been used to assess the ability of an implant to stimulate neo-osteogenesis, but, when used alone, do not allow the distinction between osteogenesis, osteoconduction, or osteoinduction. When left empty, large bony defects fill with fibrous connective tissue. The size of the defect that will not heal spontaneously must be determined experimentally for each species and for each bone. As anatomical models for craniofacial reconstruction, defects in the parietal bone were constructed to test the fate of fresh bone grafts, devitalized bone, and demineralized bone (Figure 6.1C) [583]. Full-thickness defects in cranial bone can be made with low-speed drills and ought not to cross the intraosseous sutures, which carry a risk for excess bleeding and surgical mortality [584]. A full-thickness defect in the mandibular ramus (Figure 6.1D, E) is a good site that offers an environment subject to continuous motion and significant compressive and shearing forces [585]. Non-healing intraosseous defects have been called defects in contradistinction to healing wounds; they have also been called critical size defects, the size of which varies for species, age, shape, and location [586]. Intraosseous defects can be created in small animals with drill bits or trephines.

Animal models for non-healing defects can be segmental, such as in removal of a length of a tubular bone with immobilization of the bone at either end of the defect. Alternately, they can rely on an adjacent bone for stability; for example, the radius can support a limb following ulnar osteotomy [587]. An animal's biology, anatomy, age, and the potential for spontaneous healing of the experimental defect are taken into account when designing a model for research of bone repair [588].

To model the setting for onlay bone augmentation, a clinical problem in the atrophic edentuous mandible and congenitally hypoplastic midface bones, it is possible to evaluate materials implanted in the subgingival tissue in the jaws where there are no teeth (Figure 6.1D, F) [412, 589] or in muscle overlying bony structures [583].

Segmental defects are typically created in femur, tibia, radius, ulna, or even metatarsal diaphyses, with the general rule that the length be 1.5 to 2 times the diameter of the bone (Figure 6.1G) [590]. Validation is needed with empty defects to prove that spontaneous repair is minimal in order for the model to be useful. There is always the possibility that blood can pool in a void and provide a stable clot for osteoconduction. That normal response to bone injury must be avoided if a non-healing defect is intended. Empty scaffolds must also be used as another control to show the magnitude of osteoconduction in response to the scaffold.

6.1.4 CLINICAL RELEVANCE

Connectivity is the term to describe the resemblance between an experimental model and a clinical setting. Models that are used to test a device or construct's efficacy to synthesize new bone are usually established in healthy animals. Although cost and convenience are important factors, the clinical relevance of findings from animal models is often [591]. Further, the doses of bioactive factors that are relevant in animals may be different in humans. In fact, BMP-2 stimulates osteoblast differentiation in marrow-derived osteoblast progenitor cells from rodents, but it does not do so in cells from humans [592]. Those observations were explained by studies showing that, in contrast to signaling in rodent cells, PI3-K activation is needed to regulate osteoblast differentiation in human MSC at a site downstream or independent of BMP-mediated Smad signaling [593]. Such findings lead to different considerations about clinical therapies.

Effects of tissue-engineered devices in healthy immature animals may have little relevance to intended use in aged, diseased, or chronically medicated patients. Funding agencies and researchers [594] continue to encourage research that develops or enhances model systems that better translate to clinical problems. The choice of animal and tissue models obviously depends on the question being asked and how the model relates to that question.

Strains of rodents with immune disorders are often used to evaluate cellular constructs or acellular tissue matrices from other species in order to avoid rejection; the term xenogeneic means transplantation from one species to another. It has been common to use those models for studies with cell lines established from human tumors [595]. Athymic animals, commonly the immunodeficient nu/nu T-cell-deficient mouse or the athymic rat, have been used when bone cells or engineered tissues from humans or other species were being evaluated for *in vivo* fate and function. They have also been used to test materials such as devitalized human demineralized bone, it is said, to avoid immunological reactions to foreign material [596]. When properly cleansed of cellular and tissue debris, as it must be for clinical use, human demineralized bone is osteoinductive in euthymic rats [583]. Such bone-derived materials ought to be tested in euthymic rats or mice in order to ensure proper processing. Connectivity of findings from experiments in athymic test animals clinical to use in euthymic (normal immune status) human subjects is very weak. In fact, they can result in mis-information because they do not reveal inflammatory reactions [402].

There are several approaches to testing tissue engineering devices and constructs to more relevant settings. Many hundreds of strains of naturally occurring genetic variants of mice have been characterized and are available for conditions such as type 1 diabetes, metabolic syndrome, specific tumors, hypercholesterolemia, tendency for atherosclerosis, multiple sclerosis, etc. Animals with characterized diseases can be used for therapeutic evaluations.

The rapid developments in genetic engineering and embryo manipulation technology have made transgenic and knock-out disease models an important category of animal disease models. Technologies for engineering the mouse genome, including knock-in, conditional knockout, and transgenics, have made it possible to create specific gene-sequence alterations and manipulate the levels and patterns of target-gene expression. With these techniques, researchers can take advantage of specific disease models to assess efficacy of new therapeutic modalities. Although perhaps less useful for tissue engineering research because of size, zebrafish (*Danio rerio*) models can provide diseased cells for use in gene therapy research.

Induced models of disease are those in which healthy animals are treated in some manner to recreate a setting similar to a specific disease. Few induced models completely mimic the etiology, course, and pathology of the target disease in humans, but they can be used cautiously to address specific questions. Modified diets can be very useful, but it is crucial that there be a control for different food intakes by test and control animals. Ad libitum feeding means that the animals are able to eat as much or as little as they like because an abundant supply of food is accessible. Some diets can be unappealing to the animal or reduce their ability to consume, and they can falsely give the appearance of resulting in a reduction of healing by a graft or implant. Animals in which jaws have been manipulated may have difficulty chewing food pellets; the problem is solved by provided them with milled chow and to acclimate them to it prior to intervention [372, 412, 589]. Studies can be designed to test for altered food consumption and thus whether effects are confounded by nutritional status [597]. Pair-feeding is a method to ensure equivalent caloric intakes and hence growth and healing potential in groups of animals [598]. Animals on different diets should be weighed daily to ensure similar growth trajectories. If differences are found, the animals need to be pair-fed. One way to do this is to weigh the amount of food consumed by animals fed the test diet and to provide only that amount to the control animals on the next day. Because procedures such as ovariectomy produce weight gain, it can be very important to ensure caloric equivalence to interpret results about bone growth [599]. Some diets require additional precautions. In order to test for effects of vitamin D-deficiency, for example, animals must be housed with a 12-h light/dark cycle using incandescent lighting free of ultraviolet radiation in the vitamin D3 action spectrum (290–315 nm) [600].

For studies on the use of tissue engineered constructs in aged rodents, colonies are available from the National Institute on Aging [600].

For rats and mice, there are induced models of type 1 diabetes and renal injury with similarities to human diabetic nephropathy [601]. Alloxan is a pyrimidine derivative and streptozotocin is an antimicrobial agent and both are glucose analogues with very specific toxicity for pancreatic beta cells; various protocols are available for induction of hypoinsulinemia and diabetes [602]. Metabolic models of type 2 diabetes with end-organ insulin resistance are profoundly dependent on the mouse [603]. There are always concerns about physiology and pathophysiology in laboratory rodents compared with wild ones [604].

Because aged and diabetic animals can be expensive models, it can be prudent to use approaches that more narrowly model for a specific deficiency. For a project to experimentally impair bone formation in distraction osteogenesis (DO), slow release tablets of nicotine were used to inhibit neovascularization in the DO gap [369]. That study showed that nicotine inhibits osteogenesis in a rat mandibular DO model, compared with exuberant osteogenesis and mandibular advancement in animals with the placebo tablets. The tablets are available commercially and, in this case, released nicotine for a 60-day period. Such a model of impaired neovascularization in a bone healing model can be useful for devices and constructs aiming to stimulate angiogenesis and osteogenesis.

6.2   CARTILAGE

Selecting an animal model for research on the *in vivo* mechanisms and outcomes of procedures for articular cartilage repair requires understanding of its advantages and disadvantages. The ideal model is a skeletally mature animal with anatomical and functional similarities to a human joint. Articular cartilages in animals have high chondrocyte density, with adult rabbits and mice having 10-fold and 25-fold greater cell densities than humans, respectively [605]. The thickness of articular cartilage of femoral condyles of human ranges from 2.2 to 2.4 mm [606]. The thickness of articular cartilage in the femoral condyle ranges from 30 μm in mice, 0.3 mm for rabbits, 1.0 mm for the goat, to 1.7 for sheep [607]. Inadequate subchondral bone structure and density in animals may introduce artifacts because of inadequate support of the repair site. Further, the inclusion of multiple control groups to compare treatments is not always possible because of ethical or physiological concerns of introducing simultaneous bilateral treatments. The duration of the experiment to confirm the long-term outcome of the repair therapy can be cost-prohibitive [5].

For articular cartilage, the primary clinical problems are cartilage resurfacing as a treatment for osteoarthritis (OA) and repair of focal defects. The current standard-of-care for OA is treatment of symptoms, osteotomy, or joint replacement, depending on stage of disease, and the current standard-of-care for focal defects is microfracture, against which other modalities ought to be compared.

There are many animal models of OA than can be used for tissue engineering research, including those that introduce displacement of load or joint instability, as with surgical transection of the anterior cruciate ligament in the dog [608]. In addition to the mechanical models are those that directly injure the articular cartilage, like synovial instillation of lytic enzymes [609] There are some strains of mice with spontaneous development of arthritis in specific joints, such as the STR/ort mouse model, in which osteoarthritic lesions resembling human osteoarthritis are observed in most animals by 6 months [610].

Transgenic or knockout mouse models with defects in the expression of transcription factors, MMPs, angiogenic factors, or ECM proteins have provided insight into the mechanisms that control cartilage development and, in some cases, OA pathology [611]. Mice with mutations in type IX collagen or type XI collagen develop joint disorders similar to that in humans [612]. There are many rat and mouse models of rheumatoid arthritis, but they are inflammatory or immune disorders, not often used for tissue engineering [613].

The anatomy, pathology, and state-of-the-art in tissue engineering for the temporomandibular joint (TMJ) has been detailed in this series [563]. The TMJ has unique features that need attention in experimental design. The complexity of loading mechanics on the condyle is difficult to define for humans, but much information has been gained from analyses of the articulating cartilage and subchondral bone in minipig mandibles [614]. The condylar cartilage arises from the perichondrium rather than from an endochondral process [615]. Studies from rats, monkeys, and minipigs show similarities with human condylar cartilage in its extensive capacity for growth and adaptation to changes in mechanics. For example, a reproducible model of mandibular distraction in the minipigs [616] showed adaptive responses in the condyle and in the intra-articular disc [617]. Analgesic agents have been studied in sheep TMJs [618]. Because of their size and geometry, minipigs have been used for tissue engineering approaches from TMJ disorders [619].

Surgical models for articular cartilage repair use focal full-thickness defects (osteochondral defects) or partial-thickness defects (chondral defects), and weight-bearing or partial weight-bearing sites, with rabbits and dogs as common choices. As in patients, the repair of an experimental superficial defect is extremely limited in time and extent [620]. A full-thickness defect that penetrates the subchondral bone, recognized clinically as the tidemark, is very different from the superficial defect because it exposes vessels, blood, and marrow cells that contribute to the response to injury. Formation of a fibrin clot can be followed by neo-chondrogenesis, but the response is variable and unsatisfactory, most likely due to instability of the organizing clot. Various means to ensure reproducible repair include those that involve cell transplantation [621]. Grande's study in rabbits showed that transplanted chondrocytes were largely responsible for the repair tissue that filled the experimental chondral defect [622]. Those results led to the clinical application of cell-based therapy for focal lesions [623].

The chondrogenic potential of perichondrium was demonstrated in a rabbit model in which ear perichondrium was transplanted to the eburnated surface of a patella as a full-thickness defect [624]. Neochondrogenesis was greater in younger rabbits and in older rabbits started on immediate mobilization than in older rabbits immobilized for 3 weeks. Untreated wounds or those transplanted with fascia lata showed no cartilage in this 12-month study. The regenerated neocartilage showed heterogeneity upon histological and biochemical examination. Focal fibrosis was associated with neovascularization, and hyaline cartilage was more abundant in avascular regions. This model serves as a proof-of-principle of the chondrogenic potential of perichondrium. The large supply of ear perichondrium in rabbits allowed for studies on the regulation of chondrocyte differentiation *in vitro* [624]. Human ear perichondrium has massive chondrogenic potential, as evidenced by the cauliflower ear in boxers prior to the use of protective helmets, an injury that can be created in experimental animals with deposits of blood placed under the perichondrium [625].

The use of periosteum in the treatment of full-thickness chondral defects under the influence of continuous passive motion, for example, was successful in a rabbit model [626], but its transition to humans was not fully realized [627, 628].

Subsequent research concerns the use of progenitor cells rather than mature chondrocytes for cell-based approaches to cartilage repair. Recent reviews summarize the myriad *in vitro* and *in vivo* investigations concerning tissue engineering for articular cartilage [629, 630] and for intervertebral discs [631].

We [632] and others [633] have highlighted the prevailing disadvantages to cell-based therapies. The animal models reviewed here can also be used to investigate innovative cell-free approaches for enhancing the innate regenerative capacity of skeletal tissues.

CHAPTER 7

Tissue Engineering Principles for Bone and Cartilage

Tissue engineering is a design approach that seeks to generate functional restoration of a tissue by the integration of optimized properties of component materials, growth factors, and cells. Research in skeletal tissue engineering concerns ways to translate scientific discovery to clinical treatments. Tissue engineering can address tissue restoration in two ways, cell-free and cell-based [634]. In both cases, it is necessary that the new tissue be dynamic and responsive to regulatory factors. In the cell-free approach, optimal devices are to be engineered for necessary mechanical function and integration by stimulating histogenesis and restoration of lost tissue. The foundation of the cell-free approach is to stimulate the tissue bed and adjacent tissues to initiate and maximize formation of needed tissue structures. It relies on stimulation of local cells and enhancement of the body's inherent capacity for healing; the term "regenerative medicine" conveys the idea that the tissue can heal itself, but may need assistance to optimize the innate activity of cells. A three-dimensional scaffold may serve to establish the size and shape of new tissue and to deliver specific signals to the tissue bed for cell recruitment, vascular ingrowth, and histogenesis. It may be useful to think of it as an idealized synthetic blood clot, the natural organizer of healing of small lesions [635]. In defects that have poor quality tissue, are mechanically compromised, are avascular, or are in individuals with abiding systemic problems, it may be impossible to achieve sufficient tissue regeneration with cell-free systems. Cell-based approaches are intended to overcome local deficiencies by providing cells to augment the healing response. They draw on new knowledge to enhance cellular differentiation and to regulate the organization of complex tissue architecture. Cell-based approaches may use a scaffold as a carrier to deliver cells to the defect or may entail some degree of *in vitro* generation of tissue with cells. Research with cell-based constructs are important in another, very different way. It provides information about the biology of the cells and requirements for optimal histogenesis, such as the molecular factors that regulate cell differentiation, how cells respond to mechanical signals, how the ECM organizes complex tissue morphogenesis, and what nutritional factors enhance regeneration. Information from those studies can inform cell-free approaches.

7.1 BONE

Bone tissue engineering offers alternative approaches to repair congenital or acquired osseous lesions. Bone tissue engineering aims to promote the formation of new bone and its integration with surrounding host bone. Engineered devices should promote, aid, and/or regulate filling of the defect by ingrowth of host bone (osteoconduction) or new bone formation (osteogenesis, osteoinduction). The implants may be comprised of a scaffold material alone or seeded with osteogenic cells. They may contain bioactive factors, or the cells may be genetically modified [632].

7.1.1   SCAFFOLDS

The design of scaffold materials that safely, predictably, and effectively aid the repair of osseous lesions involves decisions about many composition and design variables. Characteristics being sought in a bone tissue engineering scaffold material include biocompatibility, osteocompatibility, and osteoconductivity or osteoinductivity [632]. These combined characteristics describe a material that, upon *in vivo* implantation, releases minimal amounts of inflammatory mediators, imposes minimal hindrance to bone formation, supports the ingrowth of new bone, and supports the differentiation of cells into the osteoblastic phenotype. Once designed and manufactured, scaffolds are thoroughly tested to determine how native bone responds to them upon implantation, and to elucidate the mechanism by which they may promote bone repair (osteoconduction, mosteoinduction, and/or osseointegration, Table 4.1).

Some scaffolds are resorbable: they degrade with time and are eventually replaced with new bone. Permanent scaffolds, on the other hand, persist within the defect site and may be incorporated into the newly formed bone. Whether a scaffold is resorbable, and how quickly it degrades upon implantation depends on design parameters related to chemistry and structure, and to the characteristics of the environment, for example, presence of vasculature, mechanical loading, tissue ingrowth, enzymatic activity, pH, temperature, and ionic strength [636]. Scaffolds that degrade commensurate with formation of new bone are best for maintaining mechanical stability at the defect site [637]. Excessively rapid scaffold degradation compromises mechanical stability and may lead to high levels of inflammatory products of degradation that eventually lead to bone resorption [636].

Scaffolds may be used with or without cells, and if used with cells, they may or may not be cultured prior to implantation. Cell-free scaffolds facilitate new bone formation by cells at the defect site, and they may be thought of as "bridges" that span the defect. New bone formation advances inward from the bone/scaffold interface by migration of the host cells into the scaffold. In a successful case, good-quality bone eventually bridges the defect. The scaffold provides needed mechanical stability, a biocompatible environment that is supportive of and/or conducive to the inward migration of adjacent cells, and their formation of new bone (osteogenesis), and of vasculature (angiogenesis). When a cell-free scaffold is implanted into the defect, its surface adsorbs proteins from the surrounding tissue fluids. The degree of adsorption of proteins depends on the scaffold's chemical design, but it should be optimized, as host cells migrating into the scaffold will attach to these proteins [638].

Cell-seeded scaffolds are used when the implantation of osseous cells or tissue is desired. It is currently not well understood just how "mature" the tissue constructs should be prior to implantation. *In vitro* cultivation of cell-seeded scaffolds allows the addition of physical and chemical cues that guide cell proliferation and differentiation. Enhanced control of these cues is achieved by using bioreactor technologies.

The response of bone cells to scaffold materials is most often investigated *in vitro*, whereas the implant-host bone connection and fate of the regenerated bone are characterized through *in vivo* studies.

7.1.1.1 Design Considerations

Bone tissue engineering scaffolds must be designed with close attention to chemistry, architecture, porosity, degradation rate, and adhesiveness. These properties must promote and/or support: attachment, proliferation and migration of cells to/within the scaffold, mass transfer, formation/ingrowth of bone and of a vascular network in synchrony with the scaffold's resorption, integration of the scaffold to the host bone upon *in vivo* implantation, and mechanical strength comparable to that of native bone [639]. The requirement for similarity between the mechanical strength of the scaffold and the host bone is particularly important in the repair of defects at load-bearing sites. When scaffolds are weaker than the host bone, they may fail to support physiological loads following implantation. On the other hand, scaffolds that are stiffer than the surrounding host bone may lead to stress shielding and osteolysis [413]. Stress shielding occurs when mechanical stresses are preferentially absorbed by the stronger scaffold material (as opposed to being uniformly distributed between the scaffold and the surrounding bone). The unbalanced distribution of loads leaves the host bone without the mechanical stimulation that it needs for equilibrated bone remodeling. Bone resorption predominates, and bone formation is compromised. The net result is lysis around the implant, and therefore implant failure. In addition to all of the above requirements, practical considerations regarding ease of surgical use and implantation must be taken into account in the design of scaffolds.

A scaffold's structure determines its ability to promote bone ingrowth and/or bone formation. Structural parameters include porosity, pore size, topography, and roughness. The internal structure of scaffolds must allow for osteoblasts and other cells to migrate in, maintain appropriate spacing and orientation, and deposit their synthesized products into the construct [640]. High porosity, interconnected pore networks, a large surface-area-to-volume ratio, and a certain degree of surface roughness are structural properties beneficial to the migration and differentiation of cells, the ingrowth of new bone and a vascular network, and the mass transfer between the cells and the environment [636]. Nonetheless, excessive porosity or excessively large pores may compromise mechanical performance; the ideal pore size for bone tissue engineering scaffolds has been suggested as 200-400 μm [636], or to 500 μm [641, 642]. Smaller pores (150 μm) do not support vascularization of the construct [643]. Roughness of the scaffolds' surfaces aid fixation the host bone by supporting microscopic cell/tissue ingrowth and enhanced mineralization [644]. Nanotechnology approaches are currently investigated as a means to improve the structural surface properties of bone tissue engineering scaffolds, with the subsequent benefits to bone ingrowth [645, 646].

Fabrication technologies for bone tissue engineering scaffolds include traditional textile technologies to process polymer fibers, or mixing and molding of polymers with soluble porogen agents such as salt crystals, polymer particles, or gases. Conventional scaffold fabrication technologies can provide suboptimal control over pore connectivity, resulting in non-uniform distribution of cells, gases, and nutrients throughout the scaffold's volume [647]. Novel fabrication technologies currently under development allow more precise control over the structural characteristics of scaffolds [647]. Novel technologies include rapid prototyping, electrospinning, and surface-based methods. Rapid prototyping utilizes computer-aided design and manufacturing (CAD/CAM), which allows for the precise construction of 3-D scaffolds layer-by-layer. Examples of rapid prototyping technologies include 3D fiber deposition [648] and solid free form techniques such as 3D printing, sintering, laser ablation, and stereo lithography [649]–[651]. Electrospinning scaffold fabrication employs a high electric field between a polymer solution and a collection unit; fibers from a reservoir of the polymer solution are spun into the collector unit. Electrospinning allows for the generation of scaffolds with dimensions and topology similar to bone ECM [652]. Surface-based technologies use lithographic tools to creation stackable membranes with specific computer-designed surface topologies and resolutions on the order of 1 μm to a few nanometers. The membranes can be stacked together to form 3D scaffolds [653].

Scaffold materials for bone tissue engineering may be preformed or injectable. Preformed scaffolds are solid, with a predetermined shape; therefore, invasive surgery is necessary to implant them into osseous defects. Preformed scaffolds are advantageous because their design parameters can be strictly controlled; they are indicated for use in anatomical sites with strict requirements for load-bearing (a long-bone segment) or for specific shape (tumor resection or spinal fusion), and they may be designed specifically to fit a patient's defect using information from diagnostic imaging such as CT or MRI [654]. Injectable scaffolds harden *in situ* following injection. Their implantation may not require invasive surgery; however, due to their low mechanical strength, they are indicated for non-load bearing sites (trabecular defects) or contained defects where the structural bone is largely intact. Injectable scaffolds must be designed to allow easy *in vivo* application and hardening within a clinically acceptable time frame. Injectable scaffold materials, investigated for bone tissue engineering, include polyanhydride monomers [655], poly(propylene fumarates) (PPF) [656], lactic acid-based oligomers [657] and calcium phosphate cements, among others [658]. The main limitations of injectable scaffolds are poor mechanical strength and low porosity (reviewed in [413]). Low porosity has been addressed by adding microparticles of a fast-degrading secondary material to the scaffold's formulation [659]. Another way to increase the porosity of injectable scaffolds is to introduce gaseous bubbles during injection [413].

7.1.1.2   Types of Scaffolds

Demineralized Bone Matrix (DBM).   Demineralized bone is a processed bone-derived product that predictably induces bone formation even in non-osseous locations, a process called osteoinduction. A form of demineralized bone was used for osseous reconstruction more than 100 years ago [660]. Human demineralized bone is currently used in orthopedic, oral surgical, and other reconstructive procedures. It is obtained by acid extraction of cadaver bone, and it is readily available from tissue banks. Exhaustive criteria for donor selection and strict preparation procedures have helped address concerns regarding its risks of disease transmission and immunogenecity. In fact, when properly prepared, demineralized bone matrix supports safe and effective bone regeneration, inducing the formation of bone in non-osseous sites, a property known as osteoinductivity [394, 583]. Recent *in vivo* studies focus on the optimization of demineralized bone matrix as a carrier for factors such as BMP-2 [661], BMP-7 [662], and platelet-rich plasma [663].

Synthetic Polymers.   One of the earliest reports on the use of synthetic polymer scaffolds in a bone tissue engineering application dates back almost 50 years [664]. Currently used synthetic polymers include the poly(α-hydroxyesters) polylactic acid (PLLA), polyglycolic acid (PGA), and their co-polymer poly-lactide-co-glycolide (PLGA), as well as polycaprolactone (PCL), and polypropylene fumarate (PPF), among others.

The poly(α-hydroxyesters) PGA, PLLA, and PLGA are popular choices in bone tissue engineering, in part because they hold FDA-approval for use in absorbable sutures and bone fixation pins and screws. They are most often used to make preformed scaffolds. Their degradation rates range from months (PGA and PLGA), to years (PLLA), depending mainly on molecular weight [413, 665, 666]. One limitation is that they elicit fibrous encapsulation upon implantation [667]. The surfaces of scaffolds made from these polymers can be modified to more closely resemble that of bone – a design principle called "bioinspired." This is accomplished when immersion of the scaffold in simulated body fluid, or pre-culture with matrix secreting osteoprogenitor cells (followed by de-cellularization), results in the generation of a thin mineral layer on the scaffolds' surfaces, which enhances new bone formation [668]. Cell-free PLLA, PGA, and PLGA scaffolds have been used as carriers of osteogenic factors in *in vivo* models of bone regeneration [657, 669, 670].

The only synthetic polymer scaffold currently approved by the FDA for bone tissue engineering is made of PCL; Osteopore PCL Scaffold Bone Filler (Osteopore Inc., Milpitas, CA) is a preformed scaffold for craniofacial reconstruction. PCL degrades slowly (years), and it is ductile and flexible [671]. PCL has been used in *in vivo* studies as a carrier for the drug simvastatin in a rat cranial defect [672].

PPF is another synthetic polymer used in research on bone tissue engineering. Although mechanically weaker than PGA, PLLA , PLGA, and PCL, the osteoconductivity and the versatility (it can be used in preformed or injectable forms) of PPF make it an attractive choice [673]. PPF has been used as a carrier for osteogenic and angiogenic factors in rat calvarial defects [674].

Synthetic polymers are available with an array of structural and mechanical properties, as well as with a wide range of degradation rates following *in vivo* implantation or *in vitro* cultivation [632]. In general, increased molecular weight translates into improved mechanical strength and slower degradation [675, 676]. Synthetic polymers of different molecular weights are often blended to increase the mechanical strength while optimizing the degradation rate [672]. Synthetic polymer scaffolds may be combined with other materials in order to combine their strengths and minimize the weaknesses of the parent phases. For example, a composite of PLLA and hydroxyapatite exhibited improved strength at the implant/bone interface when compared with the PLLA implant alone [677], and a composite of polysulfone and bioactive glass showed improved bioactivity and implant-bone binding strength vs. the polysulfone alone [678]. Synthetic polymers such as PLLA, PLGA, and PGA release acidic degradation products upon implantation, but they exhibit good mechanical strength. A composite of these polymers with bioactive glass, for example, offers neutralization of the acidic degradation products, with respect to the polymers alone, and improved mechanical strength, with respect to the bioactive glass alone [639].

Natural Polymers.   Bone tissue engineering scaffolds may be made from natural polymers such as agarose, alginate, fibrin, cellulose, chitosan, chondroitin sulfate, collagen, gelatin, and hyaluronic acid. They have been reviewed with extensive references [679]. These natural polymers exhibit limited mechanical strength, which limits their use to anatomical sites without a high weight bearing requirement. Their degradation mechanisms are enzymatic and/or acid hydrolysis. Degradation rates depend on degree of cross-linking or on molecular weight, and they span from hours (non-cross-linked, low molecular weight hyaluronic acid) to permanency (highly cross-linked collagens). In order to improve their mechanical strength, natural polymers have been combined with stronger materials such as hydroxyapatite or polylactide [680, 681].

Collagen is the natural polymer most widely used in clinical osseous reconstruction. Collagen carriers for bone morphogenetic proteins were used in the treatment of tibial shaft fractures, spine fusions, and long bone non-unions [682, 683]. Collagen-ceramic (hydroxyapatite and tricalcium phosphate) composites such as Formagraft^{TM}, (Nuvasive, San Diego, CA), and OssiMend^{TM} (Collagen Matrix, Inc., Franklin Lakes, NJ) are available for spinal applications [684, 685].

Ceramics.   Ceramics have been used as skeletal prostheses for 4 decades [686]. They exhibit good biocompatibility and protein-binding affinity, features that contribute to their appeal as surgical implants and drug delivery vehicles. Although they have high compressive strength, ceramics are brittle, and, therefore, their use is limited to anatomical sites without significant load bearing [413]. The degradation rates of ceramics span from weeks to years, and in some cases (for example hydroxyapatite), they are essentially permanent (reviewed in [632]). The most common ceramics in bone tissue engineering are calcium sulfates (CS) and phosphates (CP), such as hydroxyapatite (HA) and tricalcium phosphate (TCP). Of these, CS show the most rapid degradation rates (4-12 weeks) and therefore have found a niche in dental applications and in the treatment of cysts in long bones. HA and β-TCP are approved by the FDA for use as bone substitutes, and as dental and orthopedic implants. Currently used HA and β-TCP products include injectable formulations that harden *in situ,* such as α-BSM (Etex Corp, Cambridge, MA) and Bonesource HA Cement (Stryker, Kalamazoo, MI). Although injectable calcium phosphates offer a wide range of versatility in terms of setting times, mechanical strength, and degradation rates, their low porosity limits cell migration, new bone formation, and mass transfer within the implant [641, 687]. Recent approaches to increase the porosity of injectable ceramics include the addition of fast-degrading microparticles or of a gaseous phase [688, 689].

Bioactive Glasses.   The term "bioactive" refers to the property of materials to spontaneously deposit from tissue fluids a hydroxyapatite layer on their surface. Bioactive glasses undergo step-wise surface modifications upon exposure to physiological fluids *in vivo* or *in vitro* [690, 691]. The reaction sequence is as follows: (1) cations in the bioactive glass surface exchange with hydrogen ions in the physiological solution, (2) hydrolysis introduces pores into the glass surface, (3) amorphous calcium phosphate layers precipitate onto the porous glass surface, and (4) the calcium phosphate layers crystallize into a hydroxyapatite-like mineral. These surface modifications propitiate robust bone formation. The original bioactive glass was known as Bioglass, and it was developed in the late 1960s [692]. It is also known as Bioglass 45S5 due to its composition (46.1% SiO_2, 26.9% CaO, 24.4% Na2O and 2.5% P2O5). Bioglass 45S5 promotes increased bone formation when compared with hydroxyapatite alone [693]; however, its poor mechanical properties limit its use to non-weight bearing sites. A stronger bioactive glass, apatite wollastonite ($CaSiO_3$), promotes the spontaneous formation of a calcium phosphate layer over a surface layer of SiO_2. Apatite wollastonite degrades faster, and it promotes enhanced bone formation when compared with β-TCP. It has also been used as a coating for titanium implants where it was shown to improve osseointegration and osteoconductivity [694]. Several bioactive glass products are currently available for clinical use in orthopedic applications, including Novabone (Novabone Products, LLC, Alachua, FL) and the bioactive glass/demineralized bone composite Origen DBM (Nanotherapeutics, Alachua, FL).

7.1.2   BIOACTIVE FACTORS

The environment surrounding bone formation during *in vivo* skeletal development is rich in bioactive molecules that stimulate cell proliferation, migration, differentiation, chondrogenesis, osteogenesis and/or vascularization. Bone tissue engineering approaches often attempt to mimic this milieu by introducing one of more of these bioactive molecules into the environment surrounding the formation of engineered bone. Examples of bioactive molecules commonly used in bone tissue engineering include growth and differentiation factors, adhesion peptides, and osteotrophic drugs.

7.1.2.1   Delivery Methods

Bioactive factors may be incorporated into the scaffold material's formulation to be time-released upon *in vitro* cultivation or *in vivo* implantation, or they may be synthesized and delivered into the bone defect by genetically modified cells. In the case of time-release from scaffolds, the effect of bioactive factors depends on their dose and the kinetics of their release into the defect site, both of which are determined by the scaffold's pore size and degradation rate [695]. Because of the complex tertiary structures and short biological half-lives of bioactive factors, high, supraphysiological doses are often necessary to obtain the desired osteoinductive effect [696]. These high doses may lead to undesirable bone formation outside the defect area and/or to immunological reaction, as well as increased cost. Furthermore, most current technologies for the release of bioactive factors from scaffold materials result in an initial burst of release, followed by limited release over the long term,. This can limit their therapeutic effectiveness (reviewed in [697]). To further complicate the issue, most information on dosing of bioactive factors for bone tissue engineering is derived from *in vitro* and animal studies and, therefore, may not be accurately related to the clinical requirement.

Gene therapy is a means to address the challenges associated with delivery of bioactive factors to the injured site. Cells in engineered devices may be genetically modified by *ex vivo* transfection (*in vitro* expansion and transfection followed by implantation), by direct injection of adenoviral [698] or retroviral vectors [699], or by non-viral vector delivery. Ex vivo transfection is costly, complex, and inefficient. Further, cells modified *ex vivo* exhibit compromised *in vivo* survival. Adenoviral vector delivery shows improved efficiency of genetic expression, but it is costly and introduces concerns about immunogenecity, uncontrollable gene expression, and disturbance of the host genome [698]. Non-viral delivery of plasmid DNA is the least efficient method of genetic transformation, and it is associated with transient expression and low selectivity of cell targeting; however, it is economical, relatively simple, and it introduces no immunological concerns [700]. In non-viral gene delivery, a plasmid DNA encoding for the bioactive factor of interest is introduced alone or with a carrier device into the bone defect. The plasmid DNA is non-specifically incorporated by endocytosis into host cells that infiltrate the scaffold. Uptake and transfection are followed by production of the transgene product (bioactive factor). A plasmid cDNA contained in a 3D carrier constitutes a gene activated matrix (GAM) [701]. Pioneering GAMs for bone tissue engineering were made of collagen carriers with plasmid DNAs encoding for a peptide segment of human parathyroid hormone (PTH), and they showed promising results in rat rat [702] and canine [701] models.

7.1.2.2 Bone Morphogenetic Proteins

Bone morphogenetic proteins (BMPs) were deduced and named by Urist [404], who discovered that surface-demineralized, lyophilized bone could induce bone formation *in vivo* when implanted in non-osseous sites in rabbits. Proteins known as BMPs are available as recombinant products and are used to stimulate osteogenesis [703]. As many as 20 types of human BMPs are identified, with BMP-2, BMP-4, BMP-6, and BMP-9 the most osteogenic [704]. Recombinant human (rh) forms of BMP-2 and BMP-7 are used clinically for bone reconstruction. The benefits of rhBMP-2 for the treatment of open tibial shaft fractures in humans were demonstrated in the BESST study [705]. Collagen sponges carrying rhBMP-2 are currently FDA-approved in the U.S. for the treatment of open fractures of long bones, fracture non-unions, and spinal fusions, with the name INFUSE® Bone Graft (Medtronic, Minneapolis, MN), and in Europe, they are available as InductOS™ (Wyeth, Maidenhead, UK) [682]. BMP-7, also known as osteogenic protein-1 or OP-1, is allowed for use in revision posterolateral lumbar spine fusion and long bone fracture non-union in the U.S. under the Humanitarian Device Exemptions, and it is available as OP-1 Putty (Stryker, Mahwah, NJ) [683]. Recent clinical studies suggest that the therapeutic effects of rhBMPs are comparable to autologous bone graft [706, 707]. Current efforts aims to optimize the dosing and delivery of recombinant human BMPs [708]. Delivery of BMPs into the osseous defect by gene therapy has also been investigated. Adenoviral gene transfer of BMP-4 and BMP-6 cDNAs improved ectopic bone formation in athymic rats [709, 710], but for BMP-2, the effects on bone formation depended on the dosage and timing of application in rats [711], and they were not consistent in a large animal model (sheep) where inflammation and decreased bone formation were reported [712]. Plasmid DNA encoding for BMP-4 condensed with poly(ethylenimine) was delivered within porous PLGA scaffolds and implanted in critical size defects in the craniums of rats, where they enhanced bone formation and mineralization [713]. More recently, a plasmid DNA encoding for BMP-7 and delivered in a collagen carrier induced bone formation in the paraspinal muscle of rats [714].

A problem with rhBMPs is that milligram quantities of expensive recombinant proteins are required for clinical use. In fact, it was calculated that concentrations of BMP a million time greater than that found in a human body are required for spinal arthrodesis [715]. There are technical issues about delivery and stability of rhBMPs in available carriers. Soluble forms of BMPs require an insoluble matrix for activity [582]. It is likely that a single factor is not sufficient to reproduce the osteoinductive activity of demineralized bone [716]. In native bone, BMP exists as a heterodimer composed of one chain of BMP-2 covalently linked to a chain of BMP-7 [717]. Heterodimers of recombinant BMP-2 and 7 were shown to be more potent in bone induction than were homodimers containing 2 chains of BMP-2 [718]. Important research is directed at the regulation of BMP inhibitors by injected BMP, an effect that may account for low BMP activity *in vivo* [719]. In addition, more needs to be known about the roles of BMPs in cancer [720].

7.1.2.3 Angiogenic Factors

Some bioactive factors are used with the intention of promoting vascularization of the engineered bone. Because it provides nutrient and gas exchange to the cells, vascularization is crucial for the endurance of engineered bone. In addition, vascularization provides a continuous supply of factors and proteins important to cell differentiation and function. Bioactive factors used to promote engineered tissue vascularization include vascular endothelial growth factor (VEGF), basic fibroblast growth factor (bFGF), and platelet-derived growth factor (PDGF). VEGF and bFGF stimulate endothelial cells to migrate and proliferate, thus promoting the formation of capillaries. PDGF's angiogenic effects are derived from its ability to stimulate the production of bFGF and VEGF [721]. Adenovirus gene transfer of receptor activator for nuclear factor κ B ligand (RANKL) and VEGF onto bone allograft surfaces in a mouse model was associated with improved angiogenesis and long term viability of the allograft [722]. Improved new bone formation, angiogenesis, scaffold resorption and bridging of critical sized defects in rabbits' radii were reported when collagen [723] or coralline scaffolds [724] were coated with plasmid DNA encoding for human VEGF.

7.1.2.4 Factors that Promote Cell Adhesion

Adhesive peptides can be used to promote cell adhesion and differentiation, which in turn promote osteogenesis and osseointegration. Peptides containing Arg-Gly-Asp (RGD) sequences can be immobilized within tissue engineering scaffolds, where they bind cells through integrin receptors at the cells' surfaces [725, 726]. Bone formation has been stimulated *in vivo* with RGD-modified scaffolds [727]. The osteogenic effects of RGD peptide coating are concentration- and application-dependent [728, 729]. Other adhesive factors that have been investigated with promising results include those that contain heparin- and collagen-binding motifs [730, 731].

7.1.2.5 Protein Cocktails

Protein mixtures (cocktails) can be derived from blood or tissues for bone tissue engineering, although their composition may be variable. A cocktail of enamel matrix proteins (EMD) derived from developing porcine teeth is used with reported clinical success in the treatment of osseous lesions [732, 733], and long bone defects [734]. Another bioactive factor-rich protein cocktail, platelet-rich plasma (PRP) is readily prepared from autologous whole blood in the operative setting. PRP contains a high concentration of osteogenic and angiogenic factors that stimulate the formation of new bone and vascularity, and prevent inflammation [735]. Although some report the clinical benefits of the use of PRP in osseous reconstruction [736], others report little or no efficacy [737]. These opposing reports are likely due to sample-to-sample variations [738].

7.1.2.6 Synthetic Compounds

Synthetic compounds such as simvastatin, from the statin family of drugs that trigger bone regeneration by altering bone activation-resorption-formation cycles [739] have shown potential in bone tissue engineering applications [672, 740].

The osteogenic peptide TP508, which represents a receptor-binding domain of human thrombin, has been used as an alternative to osteogenic proteins because it is less costly to produce. Thrombin activates blood platelets, promotes angiogenesis, supports inflammatory cell adhesion and increases vascular permeability. PFF scaffolds carrying TP508 exhibited improved bone formation and osseous bridging of a segmental defect in rabbit radii compared with empty PFF scaffolds [695].

7.1.3 CELLS

Studies of cells for bone tissue engineering research began with periosteal progenitor cells and differentiated osteoblasts. Those cell sources require invasive harvesting procedures and are limited in the number of cells that can be harvested. Colony forming unit fibroblasts (CFU-F) is an operational term to describe a portion of marrow cells that exhibit adherence to culture substrates *in vitro*. Subsequently, CFU-F were shown to form bone *in vivo* [741] and to have differentiation potential for multiple lineages including chondrocyte, osteoblast, adipocyte, and fibroblast [742]. CFU-F are often referred to as mesenchymal stem cells, however a better term would be marrow stromal cells (MSC). *In vitro* osteogenesis by culture-expanded MSCs has been demonstrated in monolayer culture and in 3-dimensional scaffolds. The literature contains extensive reviews of research carried out using animal and human MSCs cells to generate bone [743]–[745].

Basic and preclinical research and small clinical trials indicate the feasibility of cell-based therapies for bone and cartilage tissue engineering. Many important biological matters need to be resolved in order to advance to clinical use. More information is needed about the impact of similarities and differences in MSCs from different tissue sources [746]. Integration of the literature can be difficult due to different terminology and methodology. Terminology can be an impediment to progress. Caplan acknowledged that the name he coined in the late 1980s, "Mesenchymal Stem Cells" (MSCs), signifies a theoretical population with capacities for multilineage potential and self-renewal [747]. Because these cells do not self-renew in the unlimited sense that hematopoietic stem cells do, many investigators prefer that "MSC" denote the more descriptive, operational term, marrow stromal cell. There are other functional terms, such as Caplan's suggestion that the abbreviation MSCs should now signify Medicinal Signaling Cells, on the basis of effects of their secreted products [747]. There is terminology that denotes the tissue of origin, and there is terminology that does not explicitly denote tissue of origin (Table 7.1).

Table 7.1: Some terminology used to signify post-natal progenitor cells with potential to differentiate to osteoblast and chondrocytes.

AdMSC	adipose tissue-derived mesenchymal stem cell
ADSVF	adipose-derived stromal vascular fraction
ADSC	adipose tissue-derived mesenchymal stromal cell
AMSC	amniotic mesenchymal stem cell
ASC	adipose-derived stem cell
BMDC	bone marrow-derived cell
CFU-f	colony forming unit-fibroblast
DPSC	dental pulp-derived stem cell
MSC	marrow-derived skeletal stem cell
MSC	marrow-derived stromal cell
MSC	mesenchymal stem cell
MSC	mesenchymal stromal cell
MSF	marrow stromal fibroblast
MSPC	multipotent skeletal progenitor cell
PDLSC	periodontal ligament stem cell
PMSC	placenta-derived mesenchymal stem cell
SDSC	synovium-derived stem cell
SF-MSC	synovial fluid-mesenchymal stem cell
SHED	stem cell from human exfoliated deciduous teeth
UC-MSC	umbilical cord-mesenchymal stem cell (Wharton's Jelly)
UCSC	umbilical cord-derived stem cell
UCSSC	umbilical cord stromal stem cell

Alternatives to marrow stromal cells include adipose-derived multipotent skeletal progenitor cells (MSPCs). Large numbers of MSPCs can be easily obtained by aspiration [748], but their *in vitro* properties and *in vivo* performance have not yet been fully compared with marrow-derived cells. MSPCs may also be derived from muscle satellite cells, synovial fluid, synovium, dental pulp, periodontal ligament, amniotic fluid, umbilical cord matrix, and peripheral and cord blood. It was recently suggested that all tissues contain a population of cells that are multipotent for osteogenic, chondrogenic, myogenic, and adipogenic lineages [749]. These cells may contribute to innate wound healing, and it is likely that cells of different tissue origin vary in differentiation potential and requirements for stimulation. Given such a degree of autologous availability, once the isolation of these multipotent cells is fully optimized, and their *in vitro* and *in vivo* behavior is fully understood; the issue of cell sourcing for bone tissue engineering should be greatly simplified. Adult human fibroblasts can be redirected to a pluripotent state through ectopic expression of the multiple transcription factors, octamer-4 (OCT4), sex-determining region Y-box 2 (SOX2), and either cMYC and Krupper-like factor 4 (KLF4), or NANOG and lin-28 homolog to *C. elegans* (also known as LIN28). The efficacy and safety of adult human fibroblasts thus transformed for bone tissue engineering applications have not been evaluated [750].

Some investigations suggest that adult allogeneic MSCs may be used without eliciting rejection reactions in the recipient. After expansion *in vitro*, MSCs were shown to not elicit immune responses in tests of proliferation of allogeneic lymphocytes and did not express cell-surface HLA-class II histocompatibility antigens [751]. These properties would suggest that the MSCs became immunologically privileged and could become universal donor cells. Other studies showed that they lost immunosupressive activity and were destroyed *in vivo* [752]. Most experimental studies suggest that expanded allogeneic cells do not persist for long enough after transplantation into host bone to pose an immunological threat. Also, persistence as osteoblasts and osteocytes may not pose as damaging an immunological threat as do fresh allogeneic organ and tissue transplants. Further, it may be that the production of bone by MSCs is less important than their secretion of bioactive chemokines and cytokines into the bone regeneration microenvironment.

7.1.4   *IN VITRO* STUDIES

In vitro bone tissue engineering studies employ mature osteoblasts, osteoblast-like cell lines, or osteoblast precursors derived from bone marrow [753] to generate bone and/or to study scaffold/cell interactions. As previously mentioned, MSCs are stimulated to differentiate into osteoblasts with osteoblastogenic medium supplements such as β-glycerophosphate, dexamethasone, and ascorbic acid. Mature osteoblasts may be obtained by collagenase treatment of bone fragments or from outgrowths of cancellous bone chips. Human osteosarcoma cell lines MG63 and SaOS-2 are also employed frequently [754]. Cells are cultured within the scaffold in controlled conditions for a specified period of time. Following culture, the amount, composition and morphology of newly formed bone are typically assessed by histologic, immunolabelling, biochemical, and molecular assays, and scanning or transmission electron microscopy, among others. Osteogenic differentiation of the cultivated cells is investigated through alkaline phosphatase assays and by RNA isolation and gene expression analyses. Important *in vitro* parameters that need be optimized prior to *in vivo* investigation include the rates and byproducts of degradation of the scaffold, the kinetics of bioactive factor delivery, the stability of bioactive factors, biocompatibility, cell adhesion to the scaffold, cell proliferation, migration and differentiation within the scaffold.

A common goal of research on cell-based *in vitro* bone tissue engineering is the identification of *in vitro* conditions that optimize osteogenesis by MSCs. Cultured cells may be obtained through commercial sources or from marrow donors from which extensive clinical information is available. Regardless of the source of marrow cells, the literature shows that all marrow preparations are heterogeneous, with a wide range of osteogenic potential [755]. Cells for *in vitro* studies are combined with scaffolds and/or factors specific for a proposed application. Cell-scaffold 3D constructs are often cultivated in bioreactors that enhance bone formation by providing conditions that improve the efficiency of cell seeding into scaffolds, cell migration within the scaffolds, and improved nutrient diffusion within the scaffold. Bioreactors may also improve osteogenesis by direct mechanical stimuli such as medium perfusion [756] and/or mechanical loading [757]. Bioreactors simulating a microgravity environment were designed to provide mechanical stimulation to bone cells cultivated in polymer constructs. Although those reactors showed promising results in cartilage constructs, cells for bone constructs showed minimal osteoblastic differentiation and low alkaline phosphatase activity, low calcium deposition, and low extracellular matrix synthesis (reviewed by Wiessman et al. [758]).

Vascularization of 3D tissue engineered bone is important to overcome the diffusion limitations for nutrient/waste exchange, which limit *in vivo* cellular ingrowth, viability, and osteogenesis. Engineered bone may eventually become vascularized from ingrowth of capillaries at the periphery of the implant, but this is a lengthy process. Attempts have been made to pre-vascularize bone tissue engineered constructs prior to their seeding with osteogenic cells. Pre-implantation of cell-free scaffolds in anatomical sites with rich vascularity was investigated [759]. More recently, an arteriovenous (AV) loop was engineered inside a custom-made matrix placed in an animal's groin [760]. These efforts demonstrated that prevascularization may, in fact, reduce the available space for cell seeding, and, therefore, it may be more advantageous to promote vascularization simultaneous with bone ingrowth. Coculture of endothelial and osteoprogenitor cells resulted in engineered bone constructs with rudiment vascular networks [761].

7.1.5   *IN VIVO* STUDIES

Several animal models have been used to evaluate and compare engineered bone with normal native bone (Chapter 6). Evaluation criteria include radiodensity, integration with adjacent bone, cell differentiation, formation of woven bone at the integration site followed by remodeling into lamellar bone, mechanical strength, fracture toughness, fatigue resistance, and ECM composition, among others [762]. The most appropriate implant control is fresh autogenous bone graft because that is the clinical standard-of-care. Studies that use only empty defects for comparison are limited in interpretation because they do not account for the influence of the carrier on osteoconductction or the influence of non-specific vascular response to components of the implant. Proof-of-principle of the *in vivo* potential of MSCs-seeded scaffolds in bone tissue engineering was demonstrated ectopically [763, 764] and in non-healing defects [765, 766]. Meijer and colleagues recently raised the point that although over 300 publications demonstrated that MSCs-seeded scaffolds are a feasible technology in rodents, a strikingly smaller number of studies (fewer than 10) in the literature reported success of the technology in an osseous defect in larger animals [697].

Autologous bone marrow has been used clinically for decades because it contains cells that can become osteoblasts. Marrow can be aspirated from the iliac crest, and injected percutaneously into difficult osseous defects with success [767]. Subsequent approaches concentrated the marrow in the operating room prior to injection [768]. These techniques involve injection of the cell suspension into the lesion without scaffolds.

A few clinical reports describe bone reconstruction using expanded human cells combined with scaffolds at the time of surgery. Expanded autologous MSCs seeded in macroporous hydroxyapatite scaffolds were used in three patients with large segmental bone defects. The implants were fused to the host bone 5 to 7 months after surgery, and pain-free function was restored for 6 to 7 years [769, 770]. A case report described the successful use of multiple discs of decellularized bovine bone seeded with autologous marrow cells after 3-weeks culture in a perfusion chamber [771]. Another study employed a two-stage approach: MSCs were seeded onto a titanium/hydroxyapatite scaffold which was then placed into a patient's latissimus dorsi 7 weeks prior to transfer into a mandible segmental defect. Although a number of technical problems including fracture of the titanium mesh, infection, and necrosis were reported, areas of new bone formation and improved bone density were found within the transplant [772]. Iliac crest MSCs injected together with platelet-rich plasma around dental implants resulted in new bone formation and osseointegration [773]. Expanded periosteal cells were seeded onto a porous hydroxyapatite scaffold and used to reconstruct an avulsed phalanx; follow-up at 28 months indicated retention of the implant's volume and structure, as well as evidence of new lamellar bone within the implant [774]. In another published clinical study, patients underwent augmentation of the posterior maxilla using engineered tissue from mandibular periosteum cells seeded on a polymer scaffold and cultured for 1 week. Half the patients showed replacement with connective tissue, while the other half showed bone formation, although it was not reported whether bone formation was by the host or the donor cells [775].

Intravenous infusion of closely matched allogenic bone marrow and ablative conditioning in three infants with severe Osteogenesis Imperfecta (OI) was shown to improve growth, bone mineral content, bone histomorphometry, and reduction in fractures [776]. It was surprising that only 0 to 2.0% of osteoblasts found in these infants were of donor origin 3 months after infusion. This finding was in tandem with older studies in which donor MSCs on ceramic implants were easily found in the bone that formed over the implants' surfaces 3 weeks after implantation, but thereafter it was bone of host origin that predominated [572]. These findings raise the possibility that donor cells may not need to integrate into the new bone to attain therapeutic skeletal effects.

7.2   CARTILAGE

There are two fundamentally different types of injury to articular cartilage [777]. The Type I defect is surface damage that is restricted to the substance of articular cartilage. Because it does not involve the subchondral bone, its blood vessels, or the cellular components of marrow, the cartilage undergoes little self-restoration. The second type of articular cartilage injury, Type II, is one that penetrates into subchondral bone and can undergo satisfactory healing initially, but usually degenerates because of imperfect mechanics and variable integrity of repair. Another grading system described by Outerbridge is based on the arthroscopic appearance of chondral surface [519]. Grade I shows blistering or softening of the surface. Grade II is characterized by fissures or small clefts less than 1 cm in diameter. Grade II features deep fissures larger than 1 cm. Grade IV lesions have exposed subchondral bone.

Clinical intervention is necessary for symptomatic defects that have unstable chondral flaps, fragments, or loose bodies. Conservative therapy entails lavage with debridement and shaving, and can provide temporary relief, but without restoring damages cartilage. For advanced lesions, the aim is to generate a tissue that fills the defect, integrates with the surrounding host articular cartilage, and provides the same viscoelastic mechanical properties, as well as matrix maintenance thereafter [778]. It can be considered that marrow stimulation techniques convert Type I defects to small controlled Type II ones. Some current therapies focus on stimulating intrinsic healing, such as drilling or microfracturing of the subchondral bone plate to expose the defect to the vascular system and the bone marrow [779]. The resulting healing tissue is mainly fibrous or scar-like, as opposed to hyaline, and lacks the typical zonal organization of articular cartilage, all of which render it susceptible to degeneration [780]. Other therapies focus on transplantation of autogenous osteochondral plugs, or "mosaicplasty" [781], a technique that allows for good vertical fixation and treatment of relatively large defects, but results in donor site morbidity, chondrocyte death in the periphery of the plugs, and unpredictable integration [782]. Periosteal or perichondreal grafts are also commonly used, but although hyaline cartilage formation is reported, integration to the host site is poor [783, 784].

Autologous chondrocyte implantation (ACI) is a novel therapy that constitutes the first cell-based procedure used in the clinic for the treatment of focal cartilage defects [623]. Autologous chondrocytes are harvested and expanded *in vitro*, and then injected in suspension with fibrin beneath a periosteal flap sutured over the cartilage defect. Although concerns have been raised regarding the maintenance of the chondrocytic phenotype during *in vitro* expansion in ACI and problems with tissue integration, the technique has been used for nearly 2 decades and the results indicate that repair tissue is of varying morphology, with 78% of biopsies showing fibrocartilage after 9 to 34 months [785]. Similar clinical outcomes between ACI and microfracture have been reported 2- to 5-years after treatment [786]; 82% of ACI patients report good or excellent results after 5- to -11 years [787], and out of 45 high-level soccer players; 72% reported excellent results 41 months after ACI [788]. The method has had success in patients with early stage osteoarthritis [789]. Recently, the periosteal flap in ACI has been replaced with a collagen scaffold, in "matrix-assisted chondrocyte implantation" (MACI), which produced similar clinical outcomes to ACI after a mean follow-up of 19 months [790]. In a recent review, Khan and colleagues blamed the lack of implant-to-host cartilage integration for the eventual failure of intrinsic reparative articular cartilage processes and newly emerged surgical techniques [791].

The cellular homogeneity and avascularity of native articular cartilage render it a prototype for research on tissue engineering [778]. However, the progress of the cartilage tissue engineering field has been slower than initially expected. Limitations in proliferative capacity and synthetic activity of adult articular chondrocytes have impeded their use for engineered cartilage of clinically-relevant dimensions. Another challenge to the attainment of clinically functional engineered articular cartilage is the reproduction of the zonal organization of the native tissue. Further, different anatomical sites may require engineered cartilage of different properties. For example, the cartilage of the patellar grove experiences predominantly shear stresses *in vivo*, whereas that in the femoral condyle may experience more hydrostatic pressure [778]. Current approaches to cartilage tissue engineering aim to optimize the source of cells, the nature of the scaffold material, and the delivery of growth factors or other bioactive molecules that enhance chondrogenesis and promote the longevity and efficacy of the implant.

Although *in vitro* research for cartilage tissue engineering has yielded promising results, thus far, *in vivo* long-term hyaline-cartilage replacement products have not been achieved. Vinatier and colleagues recently attributed this lack of clinical translation to the loss of cells that occurs following transplantation, to insufficient capacity of the cells to integrate with the surrounding tissue and to induction of chondrocyte dedifferentiation by inflammatory mediators at the injury site [792].

7.2.1   SCAFFOLDS

Scaffolds for cartilage tissue engineering have many of the same requirements as those for bone tissue engineering. In particular, cartilage tissue engineering scaffolds must provide optimal mechanical durability needed for sites of weight bearing and joint loading [778].

7.2.1.1   Design Considerations

Biocompatibility, biodegradation, integration to host surrounding tissue, support of cellular viability and activity, mechanical stability, and ease of handling are desired characteristics in a cartilage tissue engineering scaffold. Although slow degradation may interfere with the deposition of new cartilage extracellular matrix, too-rapid degradation could compromise the structural support at the site of implantation.

Cartilage tissue engineering scaffolds may be hydrogels, sponges, or fibrous meshes. Hydrogels are networks made of water-insoluble natural or synthetic polymer chains that swell in water. They are highly absorbent (may contain over 99% water) and highly flexible, which allows them to fill defects of any size and shape, and be implanted with minimal invasiveness. They support transport of nutrients and waste, and allow for homogeneous suspension of cells in a 3D environment. Their mechanical properties are improved by cross-linking techniques such as photopolymerization [793]. Sponge scaffolds are structures of variable porosity, pore size, and pore interconnectivity. They are manufactured by porogen leaching, freeze-drying, or gas foaming. Mesh scaffolds are made of woven or non-woven fibers, and they may be designed with various porosities and fiber diameters. Non-woven meshes have higher porosities and surface areas that support tissue regeneration, but woven meshes afford superior mechanical strength. Mesh scaffolds may be fabricated with precise patterns using 3D fiber deposition, electrospinning. Meshes and sponges are preformed scaffolds, and as such, they may not be ideal for filling defects with irregular shapes (reviewed by Chung and Burdick [679]).

The anisotropy of articular cartilage plays a major role in its material properties and mechanical behavior, and, therefore, engineered articular cartilage of homogeneous architecture is likely to fail upon implantation. A recent area of interest is the design of scaffolds that guide engineered cartilage generation following a pattern mimetic of native tissue anisotropy. For example, hydrogel bi-layers of poly(ethylene oxide) diacrylate (PEODA) or multiple layers of poly(ethylene glycol) diacrylate (PEGDA) were seeded via sequential photopolymerization with chondrocytes isolated from the distinct articular cartilage zones. The multilayer PEGDA cell-seeded constructs had a similar cell and extracellular matrix organization to native cartilage, and the bi-layer constructs were mechanical stronger than homogeneously seeded constructs [794]. Engineered cartilage zonal organization has also been modulated via scaffold pore architecture. Poly(ethylene glycol)-terephthalate-poly(butylene terephthalate) (PEGT/PBT) composite scaffolds with pore-size gradients were developed via 3D fiber deposition. When seeded with cells, the scaffolds produced zonal distributions of GAGs and type II collagen similar to native cartilage [795].

7.2.1.2 Types of Scaffolds

Synthetic Polymers.   Scaffolds made from the poly(α-hydroxyesters) PLLA, PGA, and their copolymer PLGA, have been and continue to be widely used in cartilage tissue engineering, mostly as meshes or sponges. Although the hydrophobicity of PLGA impedes cell attachment, this limitation can be mitigated by blending with other materials such as poly(vinyl alcohol). PGA is more hydrophilic and therefore degrades more rapidly than PLLA or PLGA after *in vivo* implantation or *in vitro* culture. As previously discussed, PGA/PLLA copolymers may be optimized for mechanical and degradation properties. Poly(α-hydroxyester) scaffolds may be modified to more closely resemble the articular cartilage native environment by immobilizing biological agents on their surfaces. For example, immobilization of type II collagen onto the surface of PLLA and PLGA prevented infiltration by host tissue and capsule formation, promoted complete repair, and prevented inflammatory reaction for 6 months after *in vivo* implantation in a rabbit model [796]. Immobilization of hyaluronic acid on the surface of PLGA scaffolds increased *in vitro* attachment and chondrogenesis by bovine articular chondrocytes [797]. Incorporation of growth factors such as TGF-β1 enhances chondrogenic differentiation [798]. Synthetic polymer meshes can also be used to provide a support of predetermined shape to cells entrapped in a gel phase [799]. Poly(α-hydroxyesters) have also found a niche as materials for fabrication of microspheres loaded with growth factors and incorporated into hydrogels [800].

Other synthetic materials used in cartilage tissue engineering research include poly-caprolactone [801], poly(ethylene glycol-)terephthalate or poly(butylene terephthalate) copolymer blocks [802, 803], polyvinyl alcohol [804], oligo(poly(ethylene glycol) fumarate) [805, 806], and polypropylene fumarate [807], among others.

Natural Polymers.   Natural polymers used in cartilage tissue engineering include agarose, alginate, chitosan, collagen, hyaluronan, fibrin, and silk.

Type I collagen scaffolds seeded with chondrocytes support cartilaginous tissue formation *in vivo* [808] and *in vitro* [809]. Compared with monolayer culture, bovine articular chondrocytes maintained chondrogenesis significantly better in 3D porous collagen sponges, as assessed by sulfated GAG content, 35S-sulfate incorporation, and expression of chondrocyte signature genes [810].

Membranes made of collagens type I and III are clinically used for autologous chondrocyte implantation in Europe [811], including MACI® (Verigen, Leverkusen, Germany), Maix® (Matri-cel, Hezoenrath, Germany) and Chondro-gide® Geistlich Biomaterials, Wolhusen, Switzerland). A 5-year follow-up study on 11 patients with localized cartilage defects that underwent MACI reported improved joint function in 73% of the patients, and it concluded than MACI is a suitable but cost-intensive therapy [812]. Extensive follow-up data involving larger numbers of patients is currently not available in the literature.

Autologous bone marrow cells from a 31-year old male athlete were cultured on a gel made of modified type I collagen molecules that lack the antigenic telopeptides (Atelocollagen®, Koken Co. Ltd, Tokyo, Japan) and implanted in a medial femoral condyle defect [813, 814]. Seven months after surgery, the defect was filled with hyaline-like cartilage, and significant improvements in symptoms were reported after one year with return to normal athletic activity levels.

Hyaluronic acid has been used in non-woven meshes that are resorbable and support cell adhesion and chondrogenesis both *in vivo* [815, 816] and *in vivo* [817, 818]. It has also been used in hydrogels, which are often photopolymerized with methacrylate groups to improve stability and mechanical strength [819]. Chondrogenesis by articular and auricular chondrocytes in photocrosslinked hyaluronic acid networks has been demonstrated *in vivo* and *in vitro* [820, 821]. Cross-linked hydrogels made of hyaluronic acid alone or in composites support chondrogenesis by bone marrow-derived stromal cells *in vivo* [822]. The molecular weight and concentration of hyaluronic acid within hydrogel scaffolds can be varied to have a wide range of properties; increased macromer concentration enhances mechanical properties and increases degradation times, but it affects cell viability [823]. Hyaluronic acid added to a formulation of type I collagen sponges increased cartilage-specific matrix protein expression [824], and added onto various type of scaffolds, it has increased extracellular matrix synthesis [825, 826].

A hyaluronan-based scaffold marketed as HYAFF11® is combined with autologous chondrocytes in Hyalograft C® (Fidia Advanced Biopolymers, Abano Terme, Italy). This product has been used in patients resulting in reduced pain, improved mobility, good histological appearance, and low complication rates [827].

Alginate is a polymer derived from the cell walls of brown algae. It imbibes large amounts of water, and it can be cross-linked to form stable gels. Alginate beads are often used to provide a 3D environment for cell culture; hydrogels and beads support chondrocyte expansion, chondrocyte redifferentiation following expansion, and stem cell chondrogenic differentiation [828]–[830]. The anionic nature of alginate, however, impairs chondrocyte adhesion [831]. In order to improve chondrocyte adhesion or mechanical strength, alginate gels have been used in combination with other materials [832, 833]. The introduction of the adhesive peptide RGD as a tool to improve chondrocyte adhesiveness to alginate gels has the downside of inhibiting chondrogenesis [834]. Other limitations of alginate gels include low mechanical properties and slow degradation.

Agarose is an unbranched polysaccharide derived from the cell walls of red algae. It is liquid at high temperatures but solidifies upon cooling. It has found a niche in studies of the response of 3D chondrocyte cultures to deformational loading [835, 836] because it can transmit applied mechanical forces to cells during compression [837].

Chitosan is another polysaccharide used in cartilage tissue engineering. It is synthetically modified by deacetylation of chitin, a natural compound found in the exoskeletons of crustaceans. Chitosan is liquid at room temperature and gels at physiological temperatures. It is degraded *in vivo* by lysozymes, has great biocompatilibity, bioadhesiveness, and can be cross-linked to improve its mechanical properties [838, 839]. Chitosan can be modified with growth factors or used as part of hybrid materials to support chondrocyte phenotype in 3D cultures [840, 841].

Recent focus on silk as a scaffold biomaterial has demonstrated that it supports chondrocyte and stem cell differentiation *in vitro* [842]. Silk is not a novel biomaterial; in fact, it has been used in sutures for centuries. In its natural form, silk is composed of a filament core protein named fibroin, which is coated with gluey sericin proteins. As a tissue engineering scaffold, silk is attractive because of its biocompatibility, slow degradation, and mechanical properties [843]. Silk scaffolds supported better chondrogenesis by MSC than collagen gels [844].

Bacterial cellulose offers good material properties and degradability, but it does not promote chondrogenesis to the same degree as collagen scaffolds [845]. Although chondrogenesis has been demonstrated in non-woven cellulose meshes coated with calcium phosphate [845], others report that cellulose-based engineered cartilage constructs remain soft and devoid of native cartilage-like extracellular matrix over 4 weeks of *in vitro* cultivation, even when provided with a biomimetic coating of type II collagen [846]. Recently, an injectable self-setting cellulose-based hydrogel was used with amplified autologous rabbit nasal chondrocytes to treat an articular cartilage defect. Histological analyses revealed the formation of repair tissue with similar organization to hyaline cartilage [847].

Fibrin is a compound that results from the polymerization of fibrinogen with thrombin [848]. Fibrin can be made from autologous blood, and it tends to shrink *in vivo*. Fibrin facilitates cell-matrix interaction by integrin binding. Chondrogenesis was demonstrated in cell-seeded fibrin or fibrin-composite scaffolds *in vivo* [849, 850], and *in vitro* [799, 851]. Fibrin gels were designed to provide long-term stability and to support chondrocyte proliferation and cartilage-specific extracellular matrix deposition without undergoing deformation [852]. Fibrin is also used as a carrier for the cells in autologous chondrocyte implantation where it was reported to offer technical simplicity, minimized invasiveness, and shorter surgical times [853].

7.2.2   BIOACTIVE FACTORS

7.2.2.1   Delivery Methods

As in the case of bone tissue engineering, delivery of bioactive factors for cartilage tissue engineering may be by release from the scaffold or genetic modification of the cells. Extensive literature available on gene therapy for articular diseases such as osteoarthritis or rheumatoid arthritis provides a core of information on the types of cells, gene delivery methods, and genes that can be used in cartilage tissue engineering. Candidate cells include autologous chondrocytes, MSCs, and synovial cavity cells (reviewed in [854]). Autologous chondrocytes must be harvested from the joint and expanded *in vitro* prior to transfection because direct *in vivo* transfection is hampered by the matrix that surrounds the cells. Gene transfer to synovial cells is more advanced in clinical trials, compared with other approaches. This is likely because the synovium has a large surface area lining the internal surfaces of the joint cavity and has a large cell population. Synovial cells can be transfected by direct gene delivery or injection of transfected cells into the synovial cavity; thus gene therapy may simply consist of an intra-articular injection of gene delivery vectors or transfected cells [855]. Periosteal stem cells that were transfected *ex vivo* with BMP-7 and sonic hedgehog genes and seeded onto PGA scaffolds produced hyaline cartilage-like tissue in osteochondral defects in rabbits [856]. Reports on the use of gene therapy for cartilage tissue engineering are fairly recent and limited, but they include *in vitro* differentiation and chondrogenesis by rabbit chondrocytes on gene-activated matrices made of chitosan-gelatin entrapping plasmid DNA encoding for TGF-β1 [857] and by canine chondrocytes on a gene-activated type II collagen-glycosaminoglycan matrix entrapping plasmid DNA for IGF-I [858]. Human adipose-derived stem cells transduced with an adenovirus carrying TGF-β2 and seeded onto PLGA/alginate scaffolds produced cartilage-specific ECM *in vitro*, and *in vivo* in a rat subcutaneous pocket [859].

7.2.2.2   TGF-β

Members of the TGF-β family have been shown to enhance chondrogenic differentiation of embryonic stem cells [860] and adult MSCs [861]. In addition to their mitogenic effects [862, 863], they increase synthesis of cartilaginous extracellular matrix by cells seeded onto three-dimensional scaffolds [840]. TGF-β1 enhanced chondrogenesis and integration in a full-thickness cartilage defect in rabbits [864]. TGF-β1 incorporated into microspheres embedded into 3D scaffolds enhanced *in vitro* chondrocyte proliferation and chondrogenesis [865, 866]. When incorporated into a hyaluronic acid hydrogel containing MSC and implanted into athymic nude mice, TGF-β3 enhanced chondrogenesis and cartilage-specific gene expression [867]. TGF-β1 functions to maintain chondrogenesis and prevent cell death in synovial-derived MSC [868]. TGF-β1 in a cell-free PCL scaffold implanted subperiosteally in rabbits induced host mesenchymal stem cell recruitment and chondrogenesis [869]. BMSCs that were seeded in chitosan/collagen scaffolds and transfected with

TGF-β1 plasmids showed increased cell proliferation and type II collagen and aggrecan expression [870]. Regulation of TGF-β delivery is needed to control chondrogenesis. Direct injection of TGF-β or of transfected cells expressing TGF-β has been observed to induce osteophyte formation, swelling, and synovial hyperplasia [871].

7.2.2.3 Bone Morphogenetic Proteins

BMP-2, -4, -6, -7, -13, and -14 are capable of stimulating chondrogenic differentiation of MSCs [872]. BMP-2 enhanced *in vitro* expression of collagen type II and aggrecan by chondrocytes and MSC [873, 874]. The stimulation of chondrogenesis *in vitro* by BMP-2 was dramatically synergized by TGF-β [875]. BMP-7 stimulated proteoglycan production and prevented fibroblast infiltration into constructs implanted in mice [876]. Further, BMP-2 and -7 acted in synergy to enhance cartilage extracellular matrix production by chondrocytes in 3D culture [877]. Recombinant BMP-7 and microfracture synergistically stimulated repair of full-thickness cartilage defects in rabbits [878]. A type I collagen sponge containing plasmid DNA for BMP-2 improved healing of a rabbit articular cartilage defect [879]. Muscle-derived stem cells that were retrovirally transduced to express BMP-4 enhanced chondrogenesis and articular cartilage repair in rats [880]. Periosteal-derived stem cells that were transfected *ex vivo* with BMP-7 and seeded on PGA scaffolds showed improved healing of a full thickness cartilage defect in rabbits [856]. Nevertheless, because BMPs tend to stimulate bone formation in ectopic locations, thorough understanding of the usefulness of BMP delivery for cartilage tissue engineering is necessary.

7.2.2.4 Fibroblast Growth Factor

Proteins in the fibroblast growth factor (FGF) family play key roles in skeletal development, as well as in wound healing. FGF2 promoted proliferation of adult articular chondrocytes, as well as proliferation and chondrogenic differentiation of MSC [881]. FGF2 preserved the chondrogenic potential of chondrocytes expanded in monolayer [882], supported *in vitro* chondrogenesis by bone marrow-derived MSC from osteoarthritic patients [883], enhanced the proliferation of periosteum-derived cells from elderly patients for cartilage tissue engineering [884], and enhanced extracellular matrix accumulation in *in vitro* engineered cartilage constructs [885]. FGF-2 supported cartilage generation in a subcutaneous mice model using chondrocyte-seeded collagen sponges [886] and in a rabbit large articular cartilage defects model using a fibrin carrier [887]. Basic fibroblast-growth factor in a novel cell-free, amorphous calcium phosphate/PLLA carrier enhanced repair of rabbit osteochondral defects with hyaline-like tissue [888]. Reports indicate, however, that various FGFs may have contradicting roles in chondrogenesis; and, therefore, a more thorough understanding of the effects of various FGFs in cartilage healing is needed.

7.2.2.5 Insulin-Like-Growth-Factor

Insulin-like-growth factor I (IGF-I) mediates cartilage homeostasis by stimulating proteoglycan and type II collagen synthesis, and promoting chondrocyte survival and proliferation [889]. IGF-I induced the differentiation of MSC into chondrocytes [890]. Bovine chondrocytes genetically modified to overexpress human IGF-I seeded into polymer scaffolds were cultured *in vitro* and implanted subcutaneously in nude mice, where cartilage-specific ECM accumulation, and mechanical properties were improved by IGF-I modification [891]. A biomimetic type II collagen/glycosaminoglycan scaffold carrying plasmid cDNA for IGF-I and seeded with articular chondrocytes promoted enhanced ECM accumulation and chondrocyte differentiation compared to an IGF-I-free scaffold [858]. IGF-I acted synergistically with dynamic loading to enhance *in vitro* chondrogenesis in chondrocyte-seeded agarose gels [892]. Concomitant use of IGF-I and TGF-β worked in synergy to enhance chondrogenesis in monolayer cultures [893] but not in 3D *in vitro* or *in vivo* engineered cartilage [800, 894].

7.2.2.6   Cell Adhesive Peptides

The use of adhesion peptides such as the RGD sequence to modify the surface of cartilage tissue engineering scaffolds was reported for poly(ethylene glycol) (PEG) hydrogels, chitosan, and alginate scaffolds, with mixed results [895]. In PEG hydrogels seeded with bovine articular chondrocytes and subjected to dynamic loading, RGD acted as a mechanoreceptor, with low RGD concentrations inhibiting collagen production and higher RGD concentrations stimulating GAG synthesis [895]. RGD-modification of a poly(ethylene glycol)-diacrylate (PEGDA) hydrogels supported *in vitro* chondrocytogenic differentiation of, and chondrogenesis by human embryonic stem cell-derived cells, in contrast with unmodified PEGDA hydrogels, which did not support significant cell growth or ECM production [896]. Modified chitosan scaffolds with immobilized RGD or epidermal growth factor (EGF) on their surface, promoted improved *in vitro* proliferation and synthesis of ECM by cells from a murine chondrogenic line [897]. In contrast, RGD modification inhibited *in vitro* chondrogenesis by bovine BMSC in 3D alginate gels [834]. These observations illustrate that cell-matrix interactions regulate chondrogenesis in tissue engineering biomaterials and highlight the contrasting results that may be attained when using bioactive factors with different biomaterials.

7.2.3   CELLS

The requirements of a cell source for cartilage tissue engineering include ease of isolation and culture, capability of expansion, and generous synthesis of cartilage-specific molecules [679]. As with any other cellular therapy, autologous cells are preferred over donor cells, in order to avoid immune rejection. Given the low cellular content and activity in articular cartilage, expansion of isolated autologous articular chondrocytes is required prior to implantation. Articular chondrocytes are slow to expand *in vitro*, and they become unstable in monolayer culture where they dedifferentiate and produce a tissue of low mechanical strength. Dedifferentiated chondrocytes are fibroblast-like, with a spindle shape, increased expression of collagen type I, and decreased expression of collagen type II and proteoglycans [898]. These changes in chondrocyte phenotype occur as early as the first passage, even when the cells are encapsulated in 3D beads [899]. Approaches to mitigate *in vitro* chondrocyte dedifferentiation during expansion include culture substrates, growth factors, 3D scaffolds, bioreactors, reduced oxygen concentration, and co-culture with primary cells (reviewed in [679]).

In a recent review, Chung and Burdick provided a comprehensive summary of the literature on various sources of cells for cartilage tissue engineering [679]. The most widely investigated cells are chondrocytes and MSC.

Chondrocytes may be derived from articular cartilage, or from alternative anatomical locations such as the ear (auricular), the nose (nasoseptal), or the ribs (costal). The advantages of sourcing chondrocytes from anatomical locations other than the joints include a more active cell metabolism, as well as a reduction in donor site morbidity. Chondrocytes from different anatomical sources differ in the biochemical, physical, and mechanical properties of the extracellular matrix that they secrete both *in vitro* and *in vivo* [679]. When compared with articular cartilage, the isolation of auricular chondrocytes has a 2-fold higher yield and the cells proliferate 4 times faster. Culture in alginate beads preserves the chondrogenic potential of auricular chondrocytes, although they do secrete some elastin. The tissue produced by auricular chondrocytes *in vitro* had greater similarity to native articular cartilage after *in vivo* implantation than did the articular chondrocyte-derived counterpart [900, 901]. Chondrocytes from nasoseptal cartilage generated an ECM with high content of GAG and a high collagen II/collagen I ratio *in vitro*, and when cultivated in monolayer, they proliferated 4 times faster than did articular chondrocytes [902]. There are a number of positive reports on 3D *in vitro* cartilage tissue engineering with nasal chondrocytes [815], [903]–[906]. The rates of proliferation and matrix-deposition of nasal chondrocytes were enhanced with growth factor supplementation of serum-free medium [907, 908]. A study that compared chondrogenesis by bovine nasal, auricular, costal, and articular chondrocytes seeded on poly(L-lactide-ε-caprolactone) scaffolds and cultured for 4 weeks, demonstrated that costal chondrocytes afforded the highest collagen type II expression followed by nasoseptal, articular, and auricular chondrocytes [909]. Another study compared effects of growth factors on nasal, auricular, and costal chondrocytes, and showed that all chondrocytes were stimulated by growth factors to proliferate, secrete GAG, and express type II collagen, but that nasal and auricular chondrocytes had better ability to redifferentiation following expansion than did articular chondrocytes [910].

In most experimental cartilage tissue engineering studies, chondrocytes are harvested from young, immature animals. Clearly, the clinical setting for autologous articular cartilage replacement is better represented by aged, and osteoarthritic chondrocytes. The characterization of *in vitro* cultivation of aged or osteoarthritic cartilage has been the focus of some investigations with disappointing results in terms of cell proliferation, cartilage-specific gene expression, chondrogenic potential, and mechanical strength of secreted matrix [911]–[913]. Although these limitations may be mitigated by the use of growth factors [912, 914, 915], of 3D culture [916, 917], of periosteum-derived cells instead of chondrocytes [918], and of bioreactors [919], further research is needed to improve the potential of these cells for clinical use.

Stem cells are an alternative to autologous chondrocytes for articular cartilage tissue engineering [920, 921]. MSCs are particularly attractive because they have a capacity for *in vitro* expansion [792]. They are most often isolated from bone marrow or adipose tissue, but may also be obtained from muscle, synovium, and periosteum. Recent reports suggest that MSCs may be able to escape immune recognition and/or inhibit host defense mechanisms (reviewed in [922]). *In vitro* chondrogenesis by bone-marrow derived stromal cells was enhanced with TGF-β in various 3D scaffolds [864], [923]–[925]. Coculture of BMSC with chondrocytes increased their proliferation and chondrogenesis ability [926]. Extensive research was carried out in the past decade to characterize adipose-derived stem cells and evaluate their utility with *in vivo* models. Adipose-derived stem cells have been isolated from adult human lipoaspirates [927] and from mouse inguinal fat-pads [928]. The cells are cultured at high density in medium supplemented with transforming growth factor β1, insulin, and transferrin. Culture with TGF-β, ascorbate, and dexamethasone in a 3D environment also enhanced chondrogenic differentiation [929]. Chondrocyte differentiation of adipose-derived stem cells has been demonstrated in a variety of scaffolds [828, 930, 931]. In some reports, the chondrocyte potential of adipose-derived stem cells was inferior to that of BMSC [927, 932]. Muscle-derived stem cells have been used as cell sources and gene delivery vehicles in type I collagen scaffolds in animal models of full thickness defect repair [933]. Synovium-derived stem cells have been cultured in a variety of 3D configurations to produce cartilaginous tissue [868, 874, 934]. A study that compared chondrogenesis by MSC from 5 different tissue sources concluded that synovium-derived cells had the most chondrocyte potential [935].

The use of embryonic stem cells for cartilage tissue engineering is limited by ethical and legal issues, as well as by difficulties in controlling their pluripotency. ESCs must aggregate into embryonic bodies prior to chondrocyte differentiation. Culture of mouse ESCs with TGF-β1, BMP-2, and BMP-4 resulted in deposition of cartilage-specific proteins [860, 936]. Human ESCs produced cartilaginous matrix in RGD-modified hydrogels [896]. Co-culture with chondrocytes induced chondrocyte differentiation [937]. hESCs isolated from the dissociation of human embryonic bodies, and allowed to self-assemble *in vitro* with TGF-β3 followed by TGF-β1 and IGF-I generated cartilaginous constructs [938]. The characteristics of hESCs may be modulated with growth factor protocols towards improved chondrogenesis. The use of ESCs for articular cartilage engineering is still at a very early stage.

Skin-derived fibroblasts may be directed towards a chondrocytic phenotype under certain culture conditions. Dermal fibroblasts express cartilage-specific proteins and stain positively for GAG and collagen type II when cultured in the presence of demineralized bone [939], on aggrecan substrates with IGF-I-supplemented medium [940], or after transfection with TGF-β1 [941].

7.2.4   *IN VITRO* STUDIES

Prolonged cultivation (4-8 weeks) may be necessary to generate engineered cartilage with optimal mechanical properties and extracellular matrix content. The conditions for long term culture may be monitored and controlled in a bioreactor environment. Bioreactors are important because they enhance mass transfer between the engineered cartilage and the culture medium and optimize nutrient exchange and waste removal. Extensive research has been carried out to design bioreactors and operating conditions that optimize cell viability and chondrogenesis. A wide variety of biore-actors for cartilage tissue engineering includes traditional and modified spinner flasks, rotating wall vessels, hydrostatic pressure systems, perfusion chambers, and mechanical loading reactors [837], [942]–[947].

Because cartilage is a tissue exposed to varying levels of mechanical loading *in vivo*, a biomimetic approach to *in vitro* cultivation of engineered cartilage has been extensively explored. Bioreactors apply mechanical signals to engineered cartilage during cultivation with loading regimes such as hydrostatic or dynamic compression. Bioreactors that apply physiological levels of hydro-static pressure enhance 3D chondrogenesis [946], [948]–[952]. The effects of hydrostatic pressure highly depend on the loading regime and are subject to extensive application-specific optimization. Reduced oxygen tension in conjunction with cyclic hydrostatic pressure increased chondrogenesis both in monolayer [953] and in 3D [950]. Another form of mechanical stimulation is dynamic com-pression, which stimulated 3D *in vitro* chondrogenesis [837, 954], whereas static compression can inhibit it [955, 956]. As with hydrostatic pressure application, dynamic compression loading regimes must be optimized for every application. The loading may be applied continuously [837, 955], or applied only once [957].

Articular chondrocytes are adapted to low environmental oxygen tensions (<5%) [958]. It has been proposed that cultivation in a low oxygen tension environment may enhance chondrogenesis partially by upregulating the cellular intake of dehydroascorbate, an important source of the cofactor for collagen production, ascorbic acid [959]. There are contradictory effects of low oxygen tension during *in vitro* cultivation of 3D engineered cartilage on chondrogenesis and/or maintenance of the chondrocytic phenotype. Some report enhancements [960]–[963], and others found inhibition [964] or down-regulation of expression of collagen type II [965]. These contradictions suggest that the effects of oxygen tension are application-dependent.

Optimization of *in vitro* articular cartilage tissue engineering for different applications may involve different scaffold designs, and biochemical and mechanical cues. Research on *in vitro* cartilage tissue engineering often involves complicated experimental designs with multifactorial matrices that cannot reasonably be tested *in vitro* due to time and cost constraints. Mathematical models that quantitatively describe the dynamically changing composition, structure, and function of engineered cartilage during cultivation are therefore valuable tools to aid the development of engineered cartilage technologies by reducing time and experimental expenses. There is a current lack of mathematical models that predict various characteristics of 3D chondrogenesis (such as amount and organization of ECM deposition, cell proliferation, or mechanical strength) as a function of *in vitro* cultivation conditions [966]. The currently available mathematical models correlate the ECM composition of the engineered cartilage constructs with time [967]–[969] and nutrients, and/or growth factors [970]. Others evaluate the mechanical properties of the engineered cartilage as a function of its ECM composition or the spatial and temporal distribution of ECM in the engineered constructs with fundamental mass transport equations [971]–[973].

One area of debate is how mature should be engineered cartilage prior to *in vivo* implantation. Excessively dense engineered tissue undergoes poor nutrient transport upon implantation and poor integration [974], whereas excessively immature engineered tissue would be mechanically compromised. It has been suggested that non-weight bearing management may be beneficial for immature constructs during the rehabilitation period.

Cell seeding is a crucial step of *in vitro* cartilage tissue engineering. Successful cell seeding involves fast and highly efficient attachment of cell to scaffolds, good cell survival rates, and uniform distribution of the seeded cells through the scaffold's volume. Furthermore, a uniform spatial distribution of cells within the scaffold is important because it lays the foundation for uniform extracellular matrix deposition and hence uniform tissue growth [975]. Cell seeding of scaffolds with mixing (rather than static) conditions improves the efficiency of cell attachment, and the uniformity of cell distribution through sponge and mesh scaffolds [976]. Specific characteristics of the mixing environment, such as fluid velocities and fluid recirculation can be used to modulate the efficiency and distribution of cells within the scaffolds [975]. Uniform distribution of cells during scaffold seeding is less of an issue with hydrogel scaffolds, because the cells are uniformly incorporated into the hydrogel formulation.

Engineered cartilage generated in *in vitro* studies is most often characterized by synthesis and accumulation of GAG and type II collagen, histological appearance, and mechanical properties. Because some ECM macromolecules secreted by the cells inside the 3D engineered cartilage leak into the culture medium, the culture medium content of ECM macromolecules is also often evaluated to determine total levels of biosynthesis. For mechanical testing, engineered cartilage is most often subjected to unconfined or confined compression tests because these represent the most demanding types of loading it would experience if implanted *in vivo* [977, 978].

7.2.5   *IN VIVO* STUDIES

Most *in vivo* studies for potential cartilage tissue engineering therapies are still at the animal model stage. There is information available on the long-term outcome of autologous chondrocyte implantation with a stabilizing scaffold. BioSeed-C is a construct of autologous chondrocytes embedded in a fibrin and a stable resorbable polymer scaffold used for the treatment of clinical symptomatic focal degenerative defects of the knee, which showed implant stability and significant improvements in pain, daily activities and "knee-related quality of life" at 4-year follow-up [979]. Hyalograft C, a hyaluronan-based scaffold seeded with autologous chondrocytes and used to treat deep osteochondral defects in the knee, showed a significant improvement at 7-year follow-up in young patients with a stable joint, normal knee alignment, and an isolated chondral defects [980].

Also being evaluated clinically is a process in which autogenous chondrocytes obtained by arthroscopic biopsy are seeded into a 3-D collagen scaffold and subjected to a mechanical protocol with a bioreactor, prior to *in vivo* transplantation [981].

Animal studies show chondrogenesis by BMSC seeded on various types of scaffolds [864], [982]–[984]. Adipose-derived stem cells seeded on polymer scaffolds generated cartilage after subcutaneous implantation in small animal models [985]. Most *in vivo* studies describe short term results (2-12 weeks), but some offer more insight into the long-term outcome (>24 weeks) [986].

Chondrogenic cells derived from murine embryonic stem cells seeded into scaffolds were implanted subcutaneously in mice, where they formed cartilage-like tissue after 3-4 weeks [987].

The literature contains many investigations of chondrogenesis and/or inflammatory response following implantation of cell-seeded scaffolds on endogenous sites (subcutaneous) in mice models [983, 984], [988]–[991]. Fewer reports are available on the direct investigation of the potential of engineered cartilage constructs to repair experimentally induced articular defects in animal models [986, 992, 993].

In vivo animal studies using minimally invasive injectable gels as vehicles for cells have been carried out. One study demonstrated enhanced chondrogenesis by hyaluronan and TGF-β3 enhanced poly(ethylene) oxide diacrylate hydrogels seeded with BMSC in a mouse subcutaneous model [867].

CHAPTER 8

Perspectives

This volume describes the essential and emerging information about bone and cartilage physiology and pathology that is needed for meaningful research in skeletal tissue engineering. There are many important and interesting fundamental research questions that need to be answered in order to translate innovative tissue engineering ideas to clinical practice. Skeletal tissue engineering research is a multi-disciplinary endeavor aimed to alleviate loss of bone and cartilage tissue mass due to debilitating clinical problems, both congenital and acquired. Communication, cross-education, and collaborations among investigators with different proficiencies are necessary to overcome the hurdles that hinder discovery and clinical application of innovative approaches to skeletal deficiencies. Interactions among public, private, and governmental organizations can also expedite, prioritize, and focus initiatives. Safety and efficacy are the guiding principles for clinical therapeutics.

Developmental biology is an ancient scientific discipline. It was Aristotle who taught that the form of an animal emerges from the formless egg. Nineteenth century embryologists used improved microscopic tools to detail the development of specific structures. It was the Nobel laureate Hans Spemann whose experiments in the 1920s and 1930s showed that specific areas in the embryo could organize developing tissues when transplanted into an undifferentiated part of a second embryo; he called these areas "organizer centers" or "organizers." For example, anterior parts of an embryo tend to produce parts of the head. Recent advances in molecular methodology provide new information about gene regulation during tissue embryogenesis. Stem cell research is showing us how soluble hormones, matrix molecules, neighboring tissues, other cells in the microenvironment, and physical factors control cell fate switches. It has been proposed that there has been inadequate interaction between developmental biologists and tissue engineers [994]. Insights from classical and contemporary development biology could add important knowledge about the tissue-specific rules of construction. Comparisons of effects of growth factors and physical forces (mechanotherapy) on engineered tissues *in vitro* may inspire practical clinical progress.

It will be necessary, however, to determine whether there are limitations to emulating embryonic tissue development for biomimetic tissue engineering. Caution is advisable. For example, experimental studies indicate how skin wound healing in fetal animals is fundamentally different from post-natal scar formation, how puberty and gonadal aging influence histogenesis, and how pathological environments alter tissue remodeling. Advances from fundamental studies of differentiation, growth, postnatal tissue maintenance, and pathophysiology are likely to improve tissue engineering strategies. Awareness of the biological macro- and microenvironmental cues should help to ensure clinical successes.

It is likely that different solutions will be needed for different applications because of size, anatomical, weight-bearing, vascular, pathological, and other parameters. In the era of healthcare cost containment, there will be an advantage for simple solutions. Cell-free devices are attractive approaches, but their limitations are exemplified by the history of implants and prostheses. Analyses of retrieved oral and orthopedic implants show long-term problems as articulating surfaces wear away and as their elements degrade [995]–[997]. Thus, it is necessary to define the effects of time on new devices. There is an abiding need for research on materials that can influence bone and cartilage formation *in situ*, but there is an apparent contraction in research & development for innovative materials and scaffolds. Most of the published literature on skeletal tissue engineering involves a handful of materials available commercially. While it may seem prudent to use starting materials already approved by regulatory agencies for clinical uses, current thinking about "reparative" or "regenerative" medicine confirms the need for novel biomimetic materials that can influence bone or cartilage formation by harnessing innate pathways of histogenesis. *In vitro* research on human skeletal progenitor cells provides new information about differentiation and histogenesis that may be applicable to new cell-free strategies for regeneration. Greater understanding of chemical and mechanical factors, receptors, and signaling pathways that activate cellular activity offer an opportunity for designing smart materials, energy fields, or other tactics to regenerate skeletal tissues. Such stimuli must be rigorously analyzed for impact on the risk of malignant transformation.

The lack of common criteria and universal standards for preparation of MSC has greatly hampered progress. It is important for students to understand the significance of the different methods used to obtain cells for tissue engineering research. There is variety in the methods used to prepare, enrich, or isolate cells with skeletal potential. For example, most studies with mouse marrow and some with human marrow are based upon the liquid medium methods developed for long-term hematopoietic culture by Dexter [998] and by Greeneberger [999]. These entail the flushing of whole marrow directly into tissue culture flasks with conditions (e.g., cell density, medium, serum) that promote the formation of a confluent adherent layer of cells. For studies of the adherent stromal cells, it is preferable to remove the non-adherent cells within 1-3 days after seeding. This is done to reduce the possibility that non-adherent cells and their derivatives contaminate the preparation by attaching to the adherent cells. Even so, in addition to the stromal cells, there are mature adhesive cells in whole marrow that will be present in the adherent layer of such preparations, such as macrophages, endothelial-like cells, adipocytes, and even differentiating osteoblasts [755, 1000].

In contrast, to enrich for skeletal progenitors, marrow cells are first fractionated by centrifugation in a solution with density 1.077; in such a solution, maturing and mature granulocytes and erythrocytes collect in the pellet because of their high cytoplasmic density, whereas the small, undifferentiated progenitor cells collect at the interface because of their low density [1001]. The low-density mononuclear cells are further enriched by centrifugation in a buffer, with the platelet-containing supernatant being discarded. Because such a preparation of mononuclear cells contains non-adherent hematopoietic stem cells as well as adherent mesenchymal stem cells and other adherent cells, it is important to remove non-adherent cells 24 hours after seeding. This procedure results in the enrichment of cells with potential for skeletal differentiation. There have been attempts to isolate the multipotential cells by negative or positive selection for cell-surface markers, but there does not appear to be enough stability and specificity for those approaches to be useful [1002]. The expression of cell-surface markers has been used to characterize cell preparations, but has not been especially useful to define skeletal progenitor cells because they are population-based parameters [1003]. Wagner et al. warned that cell surface markers were not useful to discriminate between MSCs from different sources and human fibroblasts that had no differentiation potential [1004].

The relevance of troubling data from mouse studies needs to be addressed. After initially slow growth, mouse MSCs suddenly proliferated rapidly and became tumorigenic [1005]. In other studies in mice, host-derived sarcomas developed around implants with expanded MSCs [1006]. There needs to be rigorous clarification of the context in which MSCs positively or negatively modulate carcinogenesis [1007]. There is confusion about these points, especially in light of recent reports of contamination of cultures [1008].

As reviewed in Chapter 7 and elsewhere [632], there is not enough known about the immunogenicity of expanded allogeneic progenitor cells. One question is whether immunogenicity increases as the allogeneic cells differentiate and secrete donor-specific proteins. On the other hand, it may be the eventual loss of immunosuppressive activity in MSCs that results in their destruction *in vivo* [752].

Studies tracking the fate of transplanted cells and constructs suggest poor endurance *in vivo*. As reviewed in Chapter 6, data from transplantation into immunodeficient animals cannot be extrapolated to normal recipients. There needs to be resolution of the conditions in which cell-based engineered tissues survive and generate new tissue. If multipotentiality is not the key function of transplanted MSCs, but rather, as suggested by Caplan, their function is to provide signals for host cells to regenerate [747], it follows that the revised goal is to develop simpler, safer ways to provide those signals.

It is clear that research needs to continue in both cell-free and cell-based approaches. It will be necessary to develop rigorous protocols for clinical use of skeletal progenitor cells [1009, 1010]. There have been attempts to achieve consensus on release criteria by various teams and organizations, but it may be premature to do so because of insufficient knowledge. There have been proposals for standard nomenclature [1011] and minimal standards for defining skeletal progenitor cells [1012]. The work of Wagner et al., however, indicates that cell surface markers do not have the specificity to define these cells [1004]. At this time serviceable names are those that indicate species, tissue source, and lineage potential.

More research is needed to assess the quality of repaired tissues. Suggestions for evaluation criteria were made by the Functional Tissue Engineering Conference Group [594]. It was concluded that the clinical problems must drive the research questions and that, in appropriate adult animal models, comparisons need to be made with existing clinical options. In the example of focal defects in articular cartilage, the comparison should be made with microfracture techniques. Functional outcomes should include activity monitoring, mechanical assessments, quantitative histological measures of repair, and integration with advanced imaging methods. *In vitro* studies are especially valuable to define mechanisms of action and optimization of constructs. *In vivo* studies should be long enough (1 year) in order to monitor endurance of the benefit.

In conclusion, many materials, design approaches, and manufacturing methods are amenable to tissue engineering for repair and regeneration. Within the domain of skeletal tissue and organ engineering, there are likely to be multiple solutions depending on the clinical applications. Fundamental studies of differentiation, growth, and post-natal tissue maintenance will continue to inform design strategies. Advances in materials design must parallel progress in cell and tissue biology. The realizations that tissue regeneration is complex and dynamic underscore the continuing need for innovative multidisciplinary investigations, with an eye to simple and safe therapies for disabled patients.

References

[1] Laurie, SW, Kaban, LB, et al. *Donor-site morbidity after harvesting rib and iliac bone*. Plast Reconstr Surg 1984. 73:933–8. 1

[2] Lord, CF, Gebhardt, MC, et al. *Infection in bone allografts. Incidence, nature, and treatment*. J Bone Joint Surg Am 1988. 70:369–76. 1

[3] Mankin, HJ, Mow, VC, et al. *Form and function of articular cartilage*, in Orthopaedic Basic Science, S.R. Simon, Editor. 1994, American Academy of Orthopaedic Surgeons: Rosemont, IL. p. 1–44. 1, 30, 31, 32, 33, 35, 36, 37, 39, 40, 42, 66, 77

[4] Lewis, PB, McCarty, LP, 3rd, et al. *Basic science and treatment options for articular cartilage injuries*. J Orthop Sports Phys Ther 2006. 36:717–27. DOI: 10.2519/jospt.2006.2175 1

[5] Minas, T and Glowacki, J. *Cartilage repair and regeneration*, in Operative Arthroscopy, J.B. Mc Ginty, et al., Editors. 2003, Lippincott, Williams & Wilkins: Philadelphia, PA. p. 128–138. 1, 33, 36, 65, 66, 76, 93

[6] Hepenstall, B. *Fracture healing*, in Soft and Hard Tissue Repair, Biological and Clinical Aspects, T.K. Hunt, et al., Editors. 1984, Praeger: New York, NY. p. 101–142. 3, 5, 12, 13, 55, 56

[7] Pellegrino, ED and Biltz, RM. *Mineralization in the chick embryo. I. Monohydrogen phosphate and carbonate relationships during maturation of the bone crystal complex*. Calcif Tissue Res 1972. 10:128–35. DOI: 10.1007/BF02012542 3

[8] LeGeros, RZ, Calcium phosphates in oral biology and medicine. Monographs in oral science, ed. H.M. Myers. 1991, Basel, Switzerland: Karger. 201. 3

[9] Lian, JB and Stein, GS. *The cells of bone*, in Dynamics of bone and cartilage metabolism, M.J. Seibel, S.P. Robins, and J.P. Bilezikian, Editors. 1999, Academic Press: San Diego, CA. p. 165–186. 4

[10] Owen, M and Friedenstein, AJ. *Stromal stem cells: marrow-derived osteogenic precursors*. Ciba Found Symp 1988. 136:42–60. 4

[11] Zuk, PA, Zhu, M, et al. *Multilineage cells from human adipose tissue: implications for cell-based therapies*. Tissue Eng 2001. 7:211–28. DOI: 10.1089/107632701300062859 4

[12] Pate, DW, Southerland, BS, et al. *Isolation and differentiaton of mesenchymal stem cells from rabbit muscle*. Surgical Forum 1993. XLIV:587–589. 4

[13] De Bari, C, Dell'Accio, F, et al. *Multipotent mesenchymal stem cells from adult human synovial membrane*. Arthritis Rheum 2001. 44:1928–42.
DOI: 10.1002/1529-0131(200108)44:8%3C1928::AID-ART331%3E3.0.CO;2-P 4

[14] Shi, S and Gronthos, S. *Perivascular niche of postnatal mesenchymal stem cells in human bone marrow and dental pulp*. J Bone Miner Res 2003. 18:696–704.
DOI: 10.1359/jbmr.2003.18.4.696 4

[15] Trubiani, O, Di Primio, R, et al. *Morphological and cytofluorimetric analysis of adult mesenchymal stem cells expanded ex vivo from periodontal ligament*. Int J Immunopathol Pharmacol 2005. 18:213–21. 4

[16] In 't Anker, PS, Scherjon, SA, et al. *Amniotic fluid as a novel source of mesenchymal stem cells for therapeutic transplantation*. Blood 2003. 102:1548–9. DOI: 10.1182/blood-2003-04-1291 4

[17] Nakahata, T and Ogawa, M. *Hemopoietic colony-forming cells in umbilical cord blood with extensive capability to generate mono- and multipotential hemopoietic progenitors*. J Clin Invest 1982. 70:1324–8. DOI: 10.1172/JCI110734 4

[18] Bender, JG, Unverzagt, K, et al. *Phenotypic analysis and characterization of CD34+ cells from normal human bone marrow, cord blood, peripheral blood, and mobilized peripheral blood from patients undergoing autologous stem cell transplantation*. Clin Immunol Immunopathol 1994. 70:10–8. DOI: 10.1006/clin.1994.1003 4

[19] Jilka, RL, Weinstein, RS, et al. *Osteoblast programmed cell death (apoptosis): modulation by growth factors and cytokines*. J Bone Miner Res 1998. 13:793–802.
DOI: 10.1359/jbmr.1998.13.5.793 4

[20] Ham, AW. *Some histophysiological problems peculiar to calcified tissues*. J Bone Joint Surg Am 1952. 24-A-3:701–28. 5

[21] Fessler, JH, Doege, KJ, et al. *Biosynthesis of collagen*. J Cell Biochem 1985. 28:31–7.
DOI: 10.1002/jcb.240280106 8

[22] Robins, SP. *Biochemical markers of bone metabolism. CPD Bulletin*. Clin Biochem 1999. 1:116–121. 8

[23] Watts, NB. *Clinical utility of biochemical markers of bone remodeling*. Clin Chem 1999. 45:1359–68. 8

[24] Delmas, PD. *Biochemical markers for the assessment of bone turnover*, in Osteoporsis: etiology, diagnosis and management, B.L. Riggs and J. Melton III, Editors. 1995, Lippinocott-Raven: Philadelphia, PA. p. 319–333. 8

[25] Robins, SP. *An enzyme-linked immunoassay for the collagen cross-link pyridinoline.* Biochem J 1982. 207:617–20. 8

[26] Bonde, M, Qvist, P, et al. *Immunoassay for quantifying type I collagen degradation products in urine evaluated.* Clin Chem 1994. 40:2022–5. 8

[27] Hanson, DA, Weis, MA, et al. *A specific immunoassay for monitoring human bone resorption: quantitation of type I collagen cross-linked N-telopeptides in urine.* J Bone Miner Res 1992. 7:1251–8.
DOI: 10.1002/jbmr.5650071119 8

[28] Swaminathan, R. *Biochemical markers of bone turnover.* Clin Chim Acta 2001. 313:95–105. 8

[29] Glimcher, MJ, Hodge, AJ, et al. *Macromolecular Aggregation States in Relation to Mineralization: The Collagen-Hydroxyapatite System as Studied in Vitro.* Proc Natl Acad Sci U S A 1957. 43:860–7. 8

[30] Glimcher, MJ. *Mechanism of calcification: role of collagen fibrils and collagen-phosphoprotein complexes in vitro and in vivo.* Anat Rec 1989. 224:139–53. DOI: 10.1002/ar.1092240205 9

[31] Robison, R. *The Possible Significance of Hexosephosphoric Esters in Ossification.* Biochem J 1923. 17:286–93. 9

[32] glowacki, J. *Mechanisms of biomineralization,* in Disorders of bone and mineral metabolism, F.L. Coe and M.J. Favus, Editors. 2002, Lippincott Williams & Wilkins: Philadelphia. p. 227–234. 9

[33] Nelson, DG. *The influence of carbonate on the atomic structure and reactivity of hydroxyapatite.* J Dent Res 1981. 60 Spec No C:1621–9. DOI: 10.1177/00220345810600031201 9

[34] Bonar, LC, Roufosse, AH, et al. *X-ray diffraction studies of the crystallinity of bone mineral in newly synthesized and density fractionated bone.* Calcif Tissue Int 1983. 35:202–9. DOI: 10.1007/BF02405032 10

[35] Rey, C, Renugopalakrishnan, V, et al. *Fourier transform infrared spectroscopic study of the carbonate ions in bone mineral during aging.* Calcif Tissue Int 1991. 49:251–8. DOI: 10.1007/BF02556214 10

[36] Ruff, CB and Hayes, WC. *Bone-mineral content in the lower limb. Relationship to cross-sectional geometry.* J Bone Joint Surg Am 1984. 66:1024–31. 10

[37] Mongiorgi, R, Romagnoli, R, et al. *Mineral alterations in senile osteoporosis.* Biomaterials 1983. 4:192–6. DOI: 10.1016/0142-9612(83)90009-1 10, 15

[38] Burr, DB and Turner, CH. *Biomechanical measurements in age-related bone loss*, in The aging skeleton, C.J. Rosen, J. Glowacki, and J.P. Bilezikian, Editors. 1999, Academic Press: San Diego, CA. p. 301–311. 10, 15, 26

[39] Paschalis, EP, Betts, F, et al. *FTIR microspectroscopic analysis of normal human cortical and trabecular bone*. Calcif Tissue Int 1997. 61:480–6. DOI: 10.1007/s002239900371 10

[40] Ekanayake, S and Hall, BK. *The development of acellularity of the vertebral bone of the Japanese medaka, Oryzias latipes (Teleostei; Cyprinidontidae)*. J Morphol 1987. 193:253–61. DOI: 10.1002/jmor.1051930304 10

[41] Smith-Vaniz, WF, Kaufman, LS, et al. *Species-specific patterns of hyperostosis in marine teleost fishes*. Marine Biology 1995. 121:573–580. DOI: 10.1007/BF00349291 10

[42] Kim, HM, Rey, C, et al. *Isolation of calcium-phosphate crystals of bone by non-aqueous methods at low temperature*. J Bone Miner Res 1995. 10:1589–601. DOI: 10.1002/jbmr.5650101021 10

[43] Lee, DD and Glimcher, MJ. *Three-dimensional spatial relationship between the collagen fibrils and the inorganic calcium phosphate crystals of pickerel (Americanus americanus) and herring (Clupea harengus) bone*. J Mol Biol 1991. 217:487–501. DOI: 10.1016/0022-2836(91)90752-R 10

[44] Witten, PE and Huysseune, A. *A comparative view on mechanisms and functions of skeletal remodelling in teleost fish, with special emphasis on osteoclasts and their function*. Biol Rev Camb Philos Soc 2009. 84:315–46. DOI: 10.1111/j.1469-185X.2009.00077.x 10

[45] Simmons, DJ. *Calcium and skeletal tissue physiology in teleost fishes*. Clin Orthop Relat Res 1971. 76:244–80. DOI: 10.1097/00003086-197105000-00031 11

[46] Murray, MM, Mankin, HJ, et al. *The musculoskeletal system*, in The Physiologic Basis of Surgery, J.P. O'Leary, Editor. 2002, Lippincott, Williams & Wilkins: Philadelphia, PA. p. 577–595. 11, 12, 13, 16, 17, 19, 22, 35, 38, 65, 78

[47] Martin, RB, Burr, DB, et al., Skeletal tissue mechanics. 1998, New York: Springer. xiv, 392 p. 12, 13, 22, 24, 25, 27, 28, 35, 38, 39, 40, 43, 46

[48] Stover, SM, Pool, RR, et al. *Histological features of the dorsal cortex of the third metacarpal bone mid-diaphysis during postnatal growth in thoroughbred horses*. J Anat 1992. 181 (Pt 3):455–69. 12

[49] Riggs, BL, Melton Iii, LJ, 3rd, et al. *Population-based study of age and sex differences in bone volumetric density, size, geometry, and structure at different skeletal sites*. J Bone Miner Res 2004. 19:1945–54. DOI: 10.1359/jbmr.040916 13

[50] Diaz Curiel, M, Carrasco de la Pena, JL, et al. *Study of bone mineral density in lumbar spine and femoral neck in a Spanish population. Multicentre Research Project on Osteoporosis.* Osteoporos Int 1997. 7:59–64. DOI: 10.1007/BF01623462 13

[51] Lofman, O, Larsson, L, et al. *Bone mineral density in normal Swedish women.* Bone 1997. 20:167–74. 13

[52] Vega, E, Bagur, A, et al. *Densidad mineral osea en mujeres osteoporiticas y normales de Buenos Aires.* Medicina 1993. 53: 13

[53] Lehmann, R, Wapniarz, M, et al. *Dual-energy X-ray absorptiometry at the lumbar spine in German men and women: a cross-sectional study.* Calcif Tissue Int 1995. 56:350–4. DOI: 10.1007/BF00301600 13

[54] Looker, AC, Wahner, HW, et al. *Proximal femur bone mineral levels of US adults.* Osteoporos Int 1995. 5:389–409. DOI: 10.1007/BF01622262 13

[55] Karlsson, MK, Gardsell, P, et al. *Bone mineral normative data in Malmo, Sweden. Comparison with reference data and hip fracture incidence in other ethnic groups.* Acta Orthop Scand 1993. 64:168–72. DOI: 10.3109/17453679308994562 13

[56] Burger, H, van Daele, PL, et al. *The association between age and bone mineral density in men and women aged 55 years and over: the Rotterdam Study.* Bone Miner 1994. 25:1–13. DOI: 10.1016/S0169-6009(08)80203-6 13

[57] Blunt, BA, Klauber, MR, et al. *Sex differences in bone mineral density in 1653 men and women in the sixth through tenth decades of life: the Rancho Bernardo Study.* J Bone Miner Res 1994. 9:1333–8. DOI: 10.1002/jbmr.5650090903 13

[58] Rico, H, Revilla, M, et al. *Age-related differences in total and regional bone mass: a cross-sectional study with DXA in 429 normal women.* Osteoporos Int 1993. 3:154–9. DOI: 10.1007/BF01623277 13

[59] Bagur, A, Vega, E, et al. *Discrimination of total body bone mineral density measured by dexa in vertebral osteoporosis.* Calcif Tissue Int 1995. 56:263–7. DOI: 10.1007/BF00318044 13

[60] Nuti, R, Martini, G, et al. *Age-related changes of whole skeleton and body composition in healthy men.* Calcif Tissue Int 1995. 57:336–9. DOI: 10.1007/BF00302068 13

[61] Ruegsegger, P, Durand, EP, et al. *Differential effects of aging and disease on trabecular and compact bone density of the radius.* Bone 1991. 12:99–105. DOI: 10.1016/8756-3282(91)90007-6 13

[62] Compston, JE, Mellish, RW, et al. *Age-related changes in iliac crest trabecular microanatomic bone structure in man.* Bone 1987. 8:289–92. DOI: 10.1016/8756-3282(87)90004-4 13

[63] Mosekilde, L. *Age-related changes in vertebral trabecular bone architecture–assessed by a new method.* Bone 1988. 9:247–50. DOI: 10.1016/8756-3282(88)90038-5 13

[64] Brockstedt, H, Kassem, M, et al. *Age- and sex-related changes in iliac cortical bone mass and remodeling.* Bone 1993. 14:681–91. DOI: 10.1016/8756-3282(93)90092-O 13, 17

[65] Martin, RB, Pickett, JC, et al. *Studies of skeletal remodeling in aging men.* Clin Orthop Relat Res 1980. 268–82. 13

[66] Bousson, V, Bergot, C, et al. *CT of the middiaphyseal femur: cortical bone mineral density and relation to porosity.* Radiology 2000. 217:179–87. 13

[67] Poole, KE, Mayhew, PM, et al. *Changing structure of the femoral neck across the adult female lifespan.* J Bone Miner Res 2010. 25:482–91. DOI: 10.1359/jbmr.090734 14

[68] Havill, LM, Mahaney, MC, et al. *Effects of genes, sex, age, and activity on BMC, bone size, and areal and volumetric BMD.* J Bone Miner Res 2007. 22:737–46. DOI: 10.1359/jbmr.070213 14

[69] Sigurdsson, G, Aspelund, T, et al. *Increasing sex difference in bone strength in old age: The Age, Gene/Environment Susceptibility-Reykjavik study (AGES-REYKJAVIK).* Bone 2006. 39:644–51. DOI: 10.1016/j.bone.2006.03.020 14

[70] Dalzell, N, Kaptoge, S, et al. *Bone micro-architecture and determinants of strength in the radius and tibia: age-related changes in a population-based study of normal adults measured with high-resolution pQCT.* Osteoporos Int 2009. 20:1683–94. DOI: 10.1007/s00198-008-0833-6 14

[71] Knott, L and Bailey, AJ. *Collagen cross-links in mineralizing tissues: a review of their chemistry, function, and clinical relevance.* Bone 1998. 22:181–7. DOI: 10.1016/S8756-3282(97)00279-2 14

[72] Sell, DR, Nagaraj, RH, et al. *Pentosidine: a molecular marker for the cumulative damage to proteins in diabetes, aging, and uremia.* Diabetes Metab Rev 1991. 7:239–51. DOI: 10.1002/dmr.5610070404 14

[73] Odetti, P, Rossi, S, et al. *Advanced glycation end products and bone loss during aging.* Ann N Y Acad Sci 2005. 1043:710–7. DOI: 10.1196/annals.1333.082 14

[74] Valcourt, U, Merle, B, et al. *Non-enzymatic glycation of bone collagen modifies osteoclastic activity and differentiation.* J Biol Chem 2007. 282:5691–703. DOI: 10.1074/jbc.M610536200 14

[75] Sanguineti, R, Storace, D, et al. *Pentosidine effects on human osteoblasts in vitro.* Ann N Y Acad Sci 2008. 1126:166–72. DOI: 10.1196/annals.1433.044 14

[76] Yamamoto, M, Yamaguchi, T, et al. *Serum pentosidine levels are positively associated with the presence of vertebral fractures in postmenopausal women with type 2 diabetes.* J Clin Endocrinol Metab 2008. 93:1013–9. DOI: 10.1210/jc.2007-1270 14

[77] Schwartz, AV, Garnero, P, et al. *Pentosidine and increased fracture risk in older adults with type 2 diabetes.* J Clin Endocrinol Metab 2009. 94:2380–6. DOI: 10.1210/jc.2008-2498 14

[78] Shiraki, M, Kuroda, T, et al. *Nonenzymatic collagen cross-links induced by glycoxidation (pentosidine) predicts vertebral fractures.* J Bone Miner Metab 2008. 26:93–100. DOI: 10.1007/s00774-007-0784-6 14

[79] Wang, X, Shen, X, et al. *Age-related changes in the collagen network and toughness of bone.* Bone 2002. 31:1–7. DOI: 10.1016/S8756-3282(01)00697-4 14

[80] Vashishth, D, Bertholon, C, et al. *Increased non-enzymatic glycation of cancellous bone due to decrease in remodeling during alendronate therapy of osteoporotic women.* J Bone Miner Res 2008. 23:S1-S22. 14, 26

[81] Wang, X, Li, X, et al. *Age-related changes of noncalcified collagen in human cortical bone.* Ann Biomed Eng 2003. 31:1365–71. DOI: 10.1114/1.1623488 14, 26

[82] Akkus, O, Polyakova-Akkus, A, et al. *Aging of microstructural compartments in human compact bone.* J Bone Miner Res 2003. 18:1012–9. DOI: 10.1359/jbmr.2003.18.6.1012 15

[83] Grynpas, M. *Age and disease-related changes in the mineral of bone.* Calcif Tissue Int 1993. 53 Suppl 1:S57–64. DOI: 10.1007/BF01673403 15

[84] Simmons, ED, Jr., Pritzker, KP, et al. *Age-related changes in the human femoral cortex.* J Orthop Res 1991. 9:155–67. DOI: 10.1002/jor.1100090202 15

[85] Jones, SJ, Arora, M, et al. *The rate of osteoclastic destruction of calcified tissues is inversely proportional to mineral density.* Calcif Tissue Int 1995. 56:554–8. DOI: 10.1007/BF00298589 15

[86] Bailey, AJ, Sims, TJ, et al. *Age-related changes in the biochemical properties of human cancellous bone collagen: relationship to bone strength.* Calcif Tissue Int 1999. 65:203–10. DOI: 10.1007/s002239900683 15, 26, 29, 30

[87] Currey, JD. *The effects of strain rate, reconstruction and mineral content on some mechanical properties of bovine bone.* J Biomech 1975. 8:81–6. DOI: 10.1016/0021-9290(75)90046-9 15

[88] Paschalis, EP, Betts, F, et al. *FTIR microspectroscopic analysis of human iliac crest biopsies from untreated osteoporotic bone.* Calcif Tissue Int 1997. 61:487–92. DOI: 10.1007/s002239900372 15

[89] Zioupos, P and Currey, JD. *Changes in the stiffness, strength, and toughness of human cortical bone with age.* Bone 1998. 22:57–66. DOI: 10.1016/S8756-3282(97)00228-7 15, 26

[90] Wainwright, SA, Mechanical design in organisms. 1976, London: Edward Arnold. xii, 423 p. 15

[91] Mueller, KH, Trias, A, et al. *Bone density and compostiton. Age-related and pathological changes in water and mineral content.* J Bone Joint Surg Am 1966. 48:140–8. 15, 26

[92] Ding, M, Dalstra, M, et al. *Age variations in the properties of human tibial trabecular bone.* J Bone Joint Surg Br 1997. 79:995–1002. DOI: 10.1302/0301-620X.79B6.7538 15, 29

[93] McCalden, RW, McGeough, JA, et al. *Age-related changes in the tensile properties of cortical bone. The relative importance of changes in porosity, mineralization, and microstructure.* J Bone Joint Surg Am 1993. 75:1193–205. 15, 26

[94] Tonna, EA. *Electron microscopic study of bone surface changes during aging. The loss of cellular control and biofeedback.* J Gerontol 1978. 33:163–77. 15

[95] Ortner, DJ. *Aging effects on osteon remodeling.* Calcif Tissue Res 1975. 18:27–36. DOI: 10.1007/BF02546224 16

[96] Shiels, MJ, Mastro, AM, et al. *The effect of donor age on the sensitivity of osteoblasts to the proliferative effects of TGF(beta) and 1,25(OH(2)) vitamin D(3).* Life Sci 2002. 70:2967–75. DOI: 10.1016/S0024-3205(02)01548-5 16

[97] Martinez, P, Moreno, I, et al. *Changes in osteocalcin response to 1,25-dihydroxyvitamin D(3) stimulation and basal vitamin D receptor expression in human osteoblastic cells according to donor age and skeletal origin.* Bone 2001. 29:35–41. DOI: 10.1016/S8756-3282(01)00543-9 16

[98] Gazit, D, Zilberman, Y, et al. *Bone loss (osteopenia) in old male mice results from diminished activity and availability of TGF-beta.* J Cell Biochem 1998. 70:478–88. DOI: 10.1002/(SICI)1097-4644(19980915)70:4%3C478::AID-JCB5%3E3.0.CO;2-G 16

[99] Frost, HM. *In vivo osteocyte death.* J Bone Joint Surg Am 1960. 42-A:138–43. 16

[100] Wong, SY, Kariks, J, et al. *The effect of age on bone composition and viability in the femoral head.* J Bone Joint Surg Am 1985. 67:274–83. 16

[101] Vashishth, D, Verborgt, O, et al. *Decline in osteocyte lacunar density in human cortical bone is associated with accumulation of microcracks with age.* Bone 2000. 26:375–80. DOI: 10.1016/S8756-3282(00)00236-2 16

[102] Baud, CA and Auil, E. *Osteocyte differential count in normal human alveolar bone.* Acta Anat (Basel) 1971. 78:321–7. DOI: 10.1159/000143597 16

[103] Mullender, MG, van der Meer, DD, et al. *Osteocyte density changes in aging and osteoporosis.* Bone 1996. 18:109–13. DOI: 10.1016/8756-3282(95)00444-0 16

[104] Mori, S, Harruff, R, et al. *Trabecular bone volume and microdamage accumulation in the femoral heads of women with and without femoral neck fractures.* Bone 1997. 21:521–6. DOI: 10.1016/S8756-3282(97)00200-7 16, 26

[105] Dunstan, CR, Somers, NM, et al. *Osteocyte death and hip fracture.* Calcif Tissue Int 1993. 53 Suppl 1:S113–6; discussion S116-7. DOI: 10.1007/BF01673417 16

[106] Shimizu, H, Sakamoto, M, et al. *Bone resorption by isolated osteoclasts in living versus devital-ized bone: differences in mode and extent and the effects of human recombinant tissue inhibitor of metalloproteinases.* J Bone Miner Res 1990. 5:411–8. DOI: 10.1002/jbmr.5650050415 16

[107] Cao, JJ, Wronski, TJ, et al. *Aging increases stromal/osteoblastic cell-induced osteoclastogenesis and alters the osteoclast precursor pool in the mouse.* J Bone Miner Res 2005. 20:1659–68. DOI: 10.1359/JBMR.050503 16

[108] Glowacki, J. *Influence of age on human marrow.* Calcif Tissue Int 1995. 56 Suppl 1:S50–1. 16

[109] Parfitt, AM, Mathews, CH, et al. *Relationships between surface, volume, and thickness of iliac trabecular bone in aging and in osteoporosis. Implications for the microanatomic and cellular mechanisms of bone loss.* J Clin Invest 1983. 72:1396–409. DOI: 10.1172/JCI111096 17

[110] Schaffler, MB, Choi, K, et al. *Aging and matrix microdamage accumulation in human compact bone.* Bone 1995. 17:521–25. DOI: 10.1016/8756-3282(95)00370-3 17

[111] Eastell, R, Delmas, PD, et al. *Bone formation rate in older normal women: concurrent assessment with bone histomorphometry, calcium kinetics, and biochemical markers.* J Clin Endocrinol Metab 1988. 67:741–8. 17

[112] Rizzoli, R and Bonjour, R-P. *Physiology of calcium and phosphate homeostasis,* in Dynamics of bone and cartilage metabolism, M.J. Seibel, S.P. Robins, and J.P. Bilezikian, Editors. 1999, Academic Press: San Diego, CA. p. 221–258. 18

[113] Kronenberg, H and Williams, RH, Williams textbook of endocrinology. 11th ed. 2008, Philadelphia: Saunders/Elsevier. xix, 1911 p. 18

[114] Brown, EM. *Extracellular Ca2+ sensing, regulation of parathyroid cell function, and role of Ca2+ and other ions as extracellular (first) messengers.* Physiol Rev 1991. 71:371–411. 18

[115] Bullamore, JR, Wilkinson, R, et al. *Effect of age on calcium absorption.* Lancet 1970. 2:535–7. DOI: 10.1016/S0140-6736(70)91344-9 19

[116] Heaney, RP, Recker, RR, et al. *Calcium absorption in women: relationships to calcium intake, estrogen status, and age*. J Bone Miner Res 1989. 4:469–75. DOI: 10.1002/jbmr.5650040404 19

[117] Nordin, BE, Need, AG, et al. *Effect of age on calcium absorption in postmenopausal women*. Am J Clin Nutr 2004. 80:998–1002. 19

[118] Ensrud, KE, Duong, T, et al. *Low fractional calcium absorption increases the risk for hip fracture in women with low calcium intake. Study of Osteoporotic Fractures Research Group*. Ann Intern Med 2000. 132:345–53. 19

[119] Brown, EM, Gamba, G, et al. *Cloning and characterization of an extracellular Ca(2+)-sensing receptor from bovine parathyroid*. Nature 1993. 366:575–80. DOI: 10.1038/366575a0 19

[120] Fitzpatrick, LA and Bilezikian, JP. . *Parathyroid hormone: structure, function and dynamic actions*, in Dynamics of bone and cartilage metabolism, M.J. Seibel, S.P. Robins, and J.P. Bilezikian, Editors. 1999, Academic Press: San Diego, CA. p. 187–202. 19

[121] Cochran, M, Peacock, M, et al. *Renal effects of calcitonin*. Br Med J 1970. 1:135–7. DOI: 10.1136/bmj.1.5689.135 19

[122] Carmeliet, G, Verstuyif, A, et al. *The Vitamin D hormone and its nuclear receptor: genomic mechanisms involved in bone biology*, in Dynamics of bone and cartilage metabolism, M.J. Seibel, S.P. Robins, and J.P. Bilezikian, Editors. 1999, Academic Press: San Diego, CA. p. 217–227. 19

[123] Institute of Medicine (U.S.). Standing Committee on the Scientific Evaluation of Dietary Reference Intakes., Dietary reference intakes : for calcium, phosphorus, magnesium, vitamin D, and fluoride. 1997, Washington, D.C.: National Academy Press. xv, 432 p. 19

[124] Holick, MF. *Vitamin D deficiency*. N Engl J Med 2007. 357:266–81. DOI: 10.1056/NEJMra070553 19, 21

[125] *DRIs for Calcium and Vitamin D*. Published by: Institute of Medicine of the National Academies. Last Updated: 2010. Accessed on: January 3, 2011. Available at: http://iom.edu/Reports/2010/ Dietary-Reference-Intakes-for-Calcium-and-Vitamin-D/DRI-Values.aspx. 21

[126] Simon, J, Leboff, M, et al. *Fractures in the elderly and vitamin D*. J Nutr Health Aging 2002. 6:406–12. 21

[127] Eastell, R, Yergey, AL, et al. *Interrelationship among vitamin D metabolism, true calcium absorption, parathyroid function, and age in women: evidence of an age-related intestinal resistance to 1,25-dihydroxyvitamin D action*. J Bone Miner Res 1991. 6:125–32. DOI: 10.1002/jbmr.5650060205 21

[128] Adami, S, Viapiana, O, et al. *Relationship between serum parathyroid hormone, vitamin D sufficiency, age, and calcium intake.* Bone 2008. 42:267–70. DOI: 10.1016/j.bone.2007.10.003 21

[129] Bjorkman, MP, Sorva, AJ, et al. *Elevated serum parathyroid hormone predicts impaired survival prognosis in a general aged population.* Eur J Endocrinol 2008. 158:749–53. DOI: 10.1530/EJE-07-0849 21

[130] Oudshoorn, C, van der Cammen, TJ, et al. *Ageing and vitamin D deficiency: effects on calcium homeostasis and considerations for vitamin D supplementation.* Br J Nutr 2009. 101:1597–606. DOI: 10.1017/S0007114509338842 21

[131] Wang, X, Nyman, JS, et al., Fundamental biomechanics in bone tissue engineering, in Synthesis Lectures on Tissue Engineering, K.A. Athanasiou, Editor. 2009, Morgan & Claypool: San Rafael, CA. DOI: 10.2200/S00246ED1V01Y200912TIS004 21

[132] Kaplan, FS. *Form and function of bone*, in Orthopaedic Basic Science, S.R. Simon, Editor. 1994, American Academy of Orthopaedic Surgeons: Rosemont, IL. p. 127–184. 22, 25, 26, 27, 54, 64

[133] Follet, H, Boivin, G, et al. *The degree of mineralization is a determinant of bone strength: a study on human calcanei.* Bone 2004. 34:783–9. DOI: 10.1016/j.bone.2003.12.012 22

[134] Basillais, A, Bensamoun, S, et al. *Three-dimensional characterization of cortical bone microstructure by microcomputed tomography: validation with ultrasonic and microscopic measurements.* J Orthop Sci 2007. 12:141–8. DOI: 10.1007/s00776-006-1104-z 22

[135] Perilli, E, Baleani, M, et al. *Dependence of mechanical compressive strength on local variations in microarchitecture in cancellous bone of proximal human femur.* J Biomech 2008. 41:438–46. DOI: 10.1016/j.jbiomech.2007.08.003 22

[136] Follet, H, Bruyere-Garnier, K, et al. *Relationship between compressive properties of human os calcis cancellous bone and microarchitecture assessed from 2D and 3D synchrotron microtomography.* Bone 2005. 36:340–51. DOI: 10.1016/j.bone.2004.10.011 22

[137] Parfitt, AM. *Trabecular bone architecture in the pathogenesis and prevention of fracture.* Am J Med 1987. 82:68–72. DOI: 10.1016/0002-9343(87)90274-9 22

[138] Wachter, NJ, Augat, P, et al. *Predictive value of bone mineral density and morphology determined by peripheral quantitative computed tomography for cancellous bone strength of the proximal femur.* Bone 2001. 28:133–9. DOI: 10.1016/S8756-3282(00)00455-5 22

[139] Newitt, DC, Majumdar, S, et al. *In vivo assessment of architecture and micro-finite element analysis derived indices of mechanical properties of trabecular bone in the radius.* Osteoporos Int 2002. 13:6–17. DOI: 10.1007/s198-002-8332-0 22

[140] Kohles, SS and Martinez, DA. *Elastic and physicochemical relationships within cortical bone.* J Biomed Mater Res 2000. 49:479–88.
DOI: 10.1002/(SICI)1097-4636(20000315)49:4%3C479::AID-JBM6%3E3.0.CO;2-6
22

[141] Bell, KL, Loveridge, N, et al. *Regional differences in cortical porosity in the fractured femoral neck.* Bone 1999. 24:57–64. DOI: 10.1016/S8756-3282(98)00143-4 22

[142] Wachter, NJ, Krischak, GD, et al. *Correlation of bone mineral density with strength and microstructural parameters of cortical bone in vitro.* Bone 2002. 31:90–5.
DOI: 10.1016/S8756-3282(02)00779-2 24

[143] Wolff, J, Maquet, Pt, et al., The law of bone remodeling. 1986. Originally published in 1870., Berlin, Germany: Springer-Verlag. 24

[144] Kontulainen, S, Sievanen, H, et al. *Effect of long-term impact-loading on mass, size, and estimated strength of humerus and radius of female racquet-sports players: a peripheral quantitative computed tomography study between young and old starters and controls.* J Bone Miner Res 2002. 17:2281–9. DOI: 10.1359/jbmr.2002.17.12.2281 24

[145] Loomer, PM. *The impact of microgravity on bone metabolism in vitro and in vivo.* Crit Rev Oral Biol Med 2001. 12:252–61. DOI: 10.1177/10454411010120030401 24

[146] Shaw, SR, Zernicke, RF, et al. *Mechanical, morphological and biochemical adaptations of bone and muscle to hindlimb suspension and exercise.* J Biomech 1987. 20:225–34.
DOI: 10.1016/0021-9290(87)90289-2 24

[147] Loitz, BJ and Zernicke, RF. *Strenuous exercise-induced remodelling of mature bone: relationships between in vivo strains and bone mechanics.* J Exp Biol 1992. 170:1–18. 24

[148] Xie, L, Jacobson, JM, et al. *Low-level mechanical vibrations can influence bone resorption and bone formation in the growing skeleton.* Bone 2006. 39:1059–66.
DOI: 10.1016/j.bone.2006.05.012 24

[149] Pontzer, H, Lieberman, DE, et al. *Trabecular bone in the bird knee responds with high sensitivity to changes in load orientation.* J Exp Biol 2006. 209:57–65. DOI: 10.1242/jeb.01971 24

[150] Chen, HL and Gundjian, AA. *Determination of the bone-crystallites distribution function by x ray diffraction.* Med Biol Eng 1974. 12:531–6. DOI: 10.1007/BF02478612 24

[151] Cowin, SC, Bone mechanics. 1989, Boca Raton, Fla.: CTC Press. 313 p. 24, 25

[152] Reilly, DT, Burstein, AH, et al. *The elastic modulus for bone.* J Biomech 1974. 7:271–5.
DOI: 10.1016/0021-9290(74)90018-9 25

[153] George, WT and Vashishth, D. *Damage mechanisms and failure modes of cortical bone under components of physiological loading.* J Orthop Res 2005. 23:1047–53.
DOI: 10.1016/j.orthres.2005.02.008 26

[154] Martin, RB and Burr, DB, Structure, function, and adaptation of compact bone. 1989, New York: Raven Press. xii, 275 p. 26

[155] Zioupos, P, Currey, JD, et al. *The role of collagen in the declining mechanical properties of aging human cortical bone.* J Biomed Mater Res 1999. 45:108–16.
DOI: 10.1002/(SICI)1097-4636(199905)45:2%3C108::AID-JBM5%3E3.3.CO;2-1 26

[156] Eyre, DR, Dickson, IR, et al. *Collagen cross-linking in human bone and articular cartilage. Age-related changes in the content of mature hydroxypyridinium residues.* Biochem J 1988. 252:495–500. 26

[157] Dalstra, M, Huiskes, R, et al. *Mechanical and textural properties of pelvic trabecular bone.* J Biomech 1993. 26:523–35. DOI: 10.1016/0021-9290(93)90014-6 26

[158] Goulet, RW, Goldstein, SA, et al. *The relationship between the structural and orthogonal compressive properties of trabecular bone.* J Biomech 1994. 27:375–89.
DOI: 10.1016/0021-9290(94)90014-0 26

[159] Yamada, H and Evans, FG, Strength of biological materials. 1970, Baltimore,: Williams & Wilkins. x, 297 p. 27

[160] Neil, JL, Demos, TC, et al. *Tensile and compressive properties of vertebral trabecular bone.* Transactions of the orthopaedic research society 1983. 8:344. 27

[161] Kuhn, JL, Goldstein, SA, et al. *The limitations of canine trabecular bone as a model for human: a biomechanical study.* J Biomech 1989. 22:95–107. DOI: 10.1016/0021-9290(89)90032-8 27

[162] Rohl, L, Larsen, E, et al. *Tensile and compressive properties of cancellous bone.* J Biomech 1991. 24:1143–9. DOI: 10.1016/0021-9290(91)90006-9 27

[163] Rincon-Kohli, L and Zysset, PK. *Multi-axial mechanical properties of human trabecular bone.* Biomech Model Mechanobiol 2009. 8:195–208. DOI: 10.1007/s10237-008-0128-z 27

[164] Burgers, TA, Mason, J, et al. *Compressive properties of trabecular bone in the distal femur.* J Biomech 2008. 41:1077–85. DOI: 10.1016/j.jbiomech.2007.11.018 27

[165] Carter, DR and Hayes, WC. *The compressive behavior of bone as a two-phase porous structure.* J Bone Joint Surg Am 1977. 59:954–62. 27

[166] Kang, Q, An, YH, et al. *Mechanical properties and bone densities of canine trabecular bone.* J Mater Sci Mater Med 1998. 9:263–7. DOI: 10.1023/A:1008871401543 27

140   REFERENCES

[167] Nicholson, PH, Cheng, XG, et al. *Structural and material mechanical properties of human vertebral cancellous bone.* Med Eng Phys 1997. 19:729–37.
DOI: 10.1016/S1350-4533(97)00030-1 27

[168] Hernandez, CJ, Beaupre, GS, et al. *The influence of bone volume fraction and ash fraction on bone strength and modulus.* Bone 2001. 29:74–8. DOI: 10.1016/S8756-3282(01)00467-7 27

[169] Carter, DR, Schwab, GH, et al. *Tensile fracture of cancellous bone.* Acta Orthop Scand 1980. 51:733–41. DOI: 10.3109/17453678008990868 28

[170] Rice, FL, Albers, KM, et al. *Differential dependency of unmyelinated and A delta epidermal and upper dermal innervation on neurotrophins, trk receptors, and p75LNGFR.* Dev Biol 1998. 198:57–81. DOI: 10.1016/S0012-1606(98)80029-8 28

[171] Keaveny, TM, Wachtel, EF, et al. *Differences between the tensile and compressive strengths of bovine tibial trabecular bone depend on modulus.* J Biomech 1994. 27:1137–46.
DOI: 10.1016/0021-9290(94)90054-X 28

[172] Bowman, SM, Keaveny, TM, et al. *Compressive creep behavior of bovine trabecular bone.* J Biomech 1994. 27:301–10. DOI: 10.1016/0021-9290(94)90006-X 28

[173] Zilch, H, Rohlmann, A, et al. *Material properties of femoral cancellous bone in axial loading. Part II: Time dependent properties.* Arch Orthop Trauma Surg 1980. 97:257–62.
DOI: 10.1007/BF00380706

[174] Schoenfeld, CM, Lautenschlager, EP, et al. *Mechanical properties of human cancellous bone in the femoral head.* Med Biol Eng 1974. 12:313–7. DOI: 10.1007/BF02477797 28

[175] Linde, F, Norgaard, P, et al. *Mechanical properties of trabecular bone. Dependency on strain rate.* J Biomech 1991. 24:803–9. DOI: 10.1016/0021-9290(91)90062-R 28

[176] Dendorfer, S, Maier, HJ, et al. *Fatigue damage in cancellous bone: an experimental approach from continuum to micro scale.* J Mech Behav Biomed Mater 2009.
2:113–9. DOI: 10.1016/j.jmbbm.2008.03.003 28, 29

[177] Choi, K and Goldstein, SA. *A comparison of the fatigue behavior of human trabecular and cortical bone tissue.* J Biomech 1992. 25:1371–81. DOI: 10.1016/0021-9290(92)90051-2 28

[178] Haddock, SM, Yeh, OC, et al. *Similarity in the fatigue behavior of trabecular bone across site and species.* J Biomech 2004. 37:181–7. DOI: 10.1016/S0021-9290(03)00245-8 29

[179] Rapillard, L, Charlebois, M, et al. *Compressive fatigue behavior of human vertebral trabecular bone.* J Biomech 2006. 39:2133–9. DOI: 10.1016/j.jbiomech.2005.04.033 29

[180] Makiyama, AM, Vajjhala, S, et al. *Analysis of crack growth in a 3D Voronoi structure: a model for fatigue in low density trabecular bone.* J Biomech Eng 2002. 124:512–20. DOI: 10.1115/1.1503792 29

[181] Kosmopoulos, V, Schizas, C, et al. *Modeling the onset and propagation of trabecular bone microdamage during low-cycle fatigue.* J Biomech 2008. 41:515–22. DOI: 10.1016/j.jbiomech.2007.10.020 29

[182] Cauley, JA, Palermo, L, et al. *Prevalent vertebral fractures in black women and white women.* J Bone Miner Res 2008. 23:1458–67. DOI: 10.1359/jbmr.080411 29

[183] Maravic, M, Le Bihan, C, et al. *Incidence and cost of osteoporotic fractures in France during 2001. A methodological approach by the national hospital database.* Osteoporos Int 2005. 16:1475–80. DOI: 10.1007/s00198-005-2031-0

[184] Chevalley, T, Herrmann, FR, et al. *Evaluation of the age-adjusted incidence of hip fractures between urban and rural areas: the difference is not related to the prevalence of institutions for the elderly.* Osteoporos Int 2002. 13:113–8. DOI: 10.1007/s001980200002

[185] Thompson, PW, Taylor, J, et al. *The annual incidence and seasonal variation of fractures of the distal radius in men and women over 25 years in Dorset, UK.* Injury 2004. 35:462–6. DOI: 10.1016/S0020-1383(03)00117-7

[186] Vogt, MT, Cauley, JA, et al. *Distal radius fractures in older women: a 10-year follow-up study of descriptive characteristics and risk factors. The study of osteoporotic fractures.* J Am Geriatr Soc 2002. 50:97–103. DOI: 10.1046/j.1532-5415.2002.50014.x 29

[187] Cummings, SR, Black, DM, et al. *Lifetime risks of hip, Colles', or vertebral fracture and coronary heart disease among white postmenopausal women.* Arch Intern Med 1989. 149:2445–8. DOI: 10.1001/archinte.149.11.2445 29

[188] Riggs, BL, Melton, LJ, 3rd, et al. *Population-based analysis of the relationship of whole bone strength indices and fall-related loads to age- and sex-specific patterns of hip and wrist fractures.* J Bone Miner Res 2006. 21:315–23. DOI: 10.1359/JBMR.051022 29

[189] Haentjens, P, Johnell, O, et al. *Evidence from data searches and life-table analyses for gender-related differences in absolute risk of hip fracture after Colles' or spine fracture: Colles' fracture as an early and sensitive marker of skeletal fragility in white men.* J Bone Miner Res 2004. 19:1933–44. DOI: 10.1359/jbmr.040917 29

[190] McCalden, RW, McGeough, JA, et al. *Age-related changes in the compressive strength of cancellous bone. The relative importance of changes in density and trabecular architecture.* J Bone Joint Surg Am 1997. 79:421–7. 29

[191] Piersol, GA, *Normal histology, with special reference to the structure of the human body*. 12th ed. 1920, Philadelphia & London,: J. B. Lippincott company. viii, 418 p. 30

[192] Parsons, JR. *Cartilage*, in Handbook of biomaterial properties, J. Black and G. Hastings, Editors. 1998, Chapman & Hall: London. p. 41–47. 30

[193] Hall, A. *Pyshiology of cartilage*, in Sciences Basics to Orthopaedics, S. Hughes and I. McCarthy, Editors. 1998, WB Saunders Company Ltd: London, England. p. 45–69. 31, 32, 33, 35

[194] Shieh, AC and Athanasiou, KA. *Biomechanics of single zonal chondrocytes*. J Biomech 2006. 39:1595–602. DOI: 10.1016/j.jbiomech.2005.05.002 31

[195] Oegema, T and Thompson, R. *The Zone of Calcified Cartilage: Its Role in Osteoarthritis*, in Articular Cartilage and Osteoarthritis, K.E. Kuettner, et al., Editors. 1992, Raven Press: New York, NY. p. 319–331 32

[196] Hanaoka, H. *The fate of hypertrophic chondrocytes of the epiphyseal plate. An electron microscopic study*. J Bone Joint Surg Am 1976. 58:226–9. 32

[197] Zenmyo, M, Komiya, S, et al. *Morphological and biochemical evidence for apoptosis in the terminal hypertrophic chondrocytes of the growth plate*. J Pathol 1996. 180:430–3. DOI: 10.1002/(SICI)1096-9896(199612)180:4%3C430::AID-PATH691%3E3.3.CO;2-8 32

[198] Cancedda, R, Descalzi Cancedda, F, et al. *Chondrocyte differentiation*. Int Rev Cytol 1995. 159:265–358. DOI: 10.1016/S0074-7696(08)62109-9 32

[199] Zanetti, N and Solursh, M. *Effect of cell shape on cartilage differentiation*, in Cell shape : determinants, regulation, and regulatory role, W.D. Stein and F. Bronner, Editors. 1989, Academic Press: San Diego. p. 291–327. 32

[200] Siczkowski, M and Watt, FM. *Subpopulations of chondrocytes from different zones of pig articular cartilage. Isolation, growth and proteoglycan synthesis in culture*. J Cell Sci 1990. 97 (Pt 2):349–60. 32

[201] Klein, TJ, Schumacher, BL, et al. *Tissue engineering of stratified articular cartilage from chondrocyte subpopulations*. Osteoarthritis Cartilage 2003. 11:595–602. DOI: 10.1016/S1063-4584(03)00090-6 32

[202] Hu, JC and Athanasiou, KA. *Chondrocytes from different zones exhibit characteristic differences in high density culture*. Connect Tissue Res 2006. 47:133–40. DOI: 10.1080/03008200600685392 32

[203] Kvist, AJ, Nystrom, A, et al. *The major basement membrane components localize to the chondrocyte pericellular matrix–a cartilage basement membrane equivalent?* Matrix Biol 2008. 27:22–33. DOI: 10.1016/j.matbio.2007.07.007 32

[204] Benninghoff, A. *Form und Bau der Gelenkknorpel in Ihren Bezeihungen zur Junktion. Zweiter Teil. Der aufbau des Gelenkknorpels in seinen Bezeihungen zur Funktion.* Z Zellforch 1925. 2:783–862. DOI: 10.1007/BF00583443 32

[205] poole, CA. *Chondrons – The chondrocyte and its pericellular microenvironment,* in Articular Cartilage and Osteoarthritis, K.E. Kuettner, et al., Editors. 1992, Raven Press: New York, NY. p. 201–210 33

[206] Sah, RL, Kim, YJ, et al. *Biosynthetic response of cartilage explants to dynamic compression.* J Orthop Res 1989. 7:619–36. DOI: 10.1002/jor.1100070502 33

[207] Buschmann, MD, Gluzband, YA, et al. *Mechanical compression modulates matrix biosynthesis in chondrocyte/agarose culture.* J Cell Sci 1995. 108 (Pt 4):1497–508. 33

[208] Lippiello, L, Kaye, C, et al. *In vitro metabolic response of articular cartilage segments to low levels of hydrostatic pressure.* Connect Tissue Res 1985. 13:99–107. DOI: 10.3109/03008208509152388 33

[209] Nogami, H, Aoki, H, et al. *Effects of electric current on chondrogenesis in vitro.* Clin Orthop Relat Res 1982. 243–7. 33

[210] Chandrasekhar, S and Harvey, AK. *Transforming growth factor-beta is a potent inhibitor of IL-1 induced protease activity and cartilage proteoglycan degradation.* Biochem Biophys Res Commun 1988. 157:1352–9. DOI: 10.1016/S0006-291X(88)81024-6 34

[211] Morales, TI and Roberts, AB. *Transforming growth factor beta regulates the metabolism of proteoglycans in bovine cartilage organ cultures.* J Biol Chem 1988. 263:12828–31.

[212] Hardingham, TE, Bayliss, MT, et al. *Effects of growth factors and cytokines on proteoglycan turnover in articular cartilage.* Br J Rheumatol 1992. 31 Suppl 1:1–6. 34

[213] Schmidt, TA, Gastelum, NS, et al. *Differential regulation of proteoglycan 4 metabolism in cartilage by IL-1alpha, IGF-I, and TGF-beta1.* Osteoarthritis Cartilage 2008. 16:90–7. DOI: 10.1016/j.joca.2007.05.009 34

[214] van Beuningen, HM, van der Kraan, PM, et al. *Protection from interleukin 1 induced destruction of articular cartilage by transforming growth factor beta: studies in anatomically intact cartilage in vitro and in vivo.* Ann Rheum Dis 1993. 52:185–91. DOI: 10.1136/ard.52.3.185

[215] Morales, TI and Roberts, AB. *The interaction between retinoic acid and the transforming growth factors-beta in calf articular cartilage organ cultures.* Arch Biochem Biophys 1992. 293:79–84. DOI: 10.1016/0003-9861(92)90368-7 34

[216] Tyler, JA. *Insulin-like growth factor 1 can decrease degradation and promote synthesis of proteo-glycan in cartilage exposed to cytokines.* Biochem J 1989. 260:543–8. 34

[217] Xu, C, Oyajobi, BO, et al. *Effects of growth factors and interleukin-1 alpha on proteoglycan and type II collagen turnover in bovine nasal and articular chondrocyte pellet cultures.* Endocrinology 1996. 137:3557–65. DOI: 10.1210/en.137.8.3557 34

[218] Trippel, SB. *Growth factor actions on articular cartilage.* J Rheumatol Suppl 1995. 43:129–32. 34

[219] Chandrasekhar, S and Harvey, AK. *Induction of interleukin-1 receptors on chondrocytes by fibroblast growth factor: a possible mechanism for modulation of interleukin-1 activity.* J Cell Physiol 1989. 138:236–46. DOI: 10.1002/jcp.1041380204

[220] Sonal, D. *Prevention of IGF-1 and TGFbeta stimulated type II collagen and decorin expression by bFGF and identification of IGF-1 mRNA transcripts in articular chondrocytes.* Matrix Biol 2001. 20:233–42. DOI: 10.1016/S0945-053X(01)00140-8

[221] Sah, RL, Chen, AC, et al. *Differential effects of bFGF and IGF-I on matrix metabolism in calf and adult bovine cartilage explants.* Arch Biochem Biophys 1994. 308:137–47. DOI: 10.1006/abbi.1994.1020 34

[222] Krakauer, T, Oppenheim, JJ, et al. *Human interleukin 1 mediates cartilage matrix degradation.* Cell Immunol 1985. 91:92–9. DOI: 10.1016/0008-8749(85)90034-6 34

[223] Benton, HP and Tyler, JA. *Inhibition of cartilage proteoglycan synthesis by interleukin I.* Biochem Biophys Res Commun 1988. 154:421–8. DOI: 10.1016/0006-291X(88)90703-6 34

[224] Roman-Blas, JA, Stokes, DG, et al. *Modulation of TGF-beta signaling by proinflammatory cytokines in articular chondrocytes.* Osteoarthritis Cartilage 2007. 15:1367–77. DOI: 10.1016/j.joca.2007.04.011 34

[225] Jubb, RW. *Differential responses of human articular cartilage to retinol.* Ann Rheum Dis 1984. 43:833–40. DOI: 10.1136/ard.43.6.833 34

[226] Dingle, JT, Horsfield, P, et al. *Breakdown of proteoglycan and collagen induced in pig articular cartilage in organ culture.* Ann Rheum Dis 1975. 34:303–11. DOI: 10.1136/ard.34.4.303 34

[227] Von den Hoff, HW, van Kampen, GP, et al. *Proteoglycan depletion of intact articular cartilage by retinoic acid is irreversible and involves loss of hyaluronate.* Osteoarthritis Cartilage 1993. 1:157–66. DOI: 10.1016/S1063-4584(05)80086-X 34

[228] Ilic, MZ, Mok, MT, et al. *Catabolism of aggrecan by explant cultures of human artic-ular cartilage in the presence of retinoic acid.* Arch Biochem Biophys 1995. 322:22–30. DOI: 10.1006/abbi.1995.1431 34

[229] Aigner, T, Hemmel, M, et al. *Apoptotic cell death is not a widespread phenomenon in normal aging and osteoarthritis human articular knee cartilage: a study of proliferation, programmed cell death (apoptosis), and viability of chondrocytes in normal and osteoarthritic human knee cartilage.* Arthritis Rheum 2001. 44:1304–12.
DOI: 10.1002/1529-0131(200106)44:6%3C1304::AID-ART222%3E3.0.CO;2-T 33

[230] Goldring, MB and Goldring, SR. *Osteoarthritis.* J Cell Physiol 2007. 213:626–34. DOI: 10.1002/jcp.21258 33

[231] Horton, WE, Jr., Yagi, R, et al. *Overview of studies comparing human normal cartilage with minimal and advanced osteoarthritic cartilage.* Clin Exp Rheumatol 2005. 23:103–12. 33

[232] Sandy, JD, Adams, ME, et al. *In vivo and in vitro stimulation of chondrocyte biosynthetic activity in early experimental osteoarthritis.* Arthritis Rheum 1984. 27:388–97.
DOI: 10.1002/art.1780270405 33

[233] Aigner, T, Soder, S, et al. *Mechanisms of disease: role of chondrocytes in the pathogenesis of osteoarthritis–structure, chaos and senescence.* Nat Clin Pract Rheumatol 2007. 3:391–9.
DOI: 10.1038/ncprheum0534 34

[234] Eid, K, Thornhill, TS, et al. *Chondrocyte gene expression in osteoarthritis: Correlation with disease severity.* J Orthop Res 2006. 24:1062–8. DOI: 10.1002/jor.20137 34, 78

[235] Tchetina, EV, Squires, G, et al. *Increased type II collagen degradation and very early focal cartilage degeneration is associated with upregulation of chondrocyte differentiation related genes in early human articular cartilage lesions.* J Rheumatol 2005. 32:876–86. 34

[236] Linn, FC and Sokoloff, L. *Movement and Composition of Interstitial Fluid of Cartilage.* Arthritis Rheum 1965. 8:481–94. DOI: 10.1002/art.1780080402 35, 39

[237] Lai, WM, Hou, JS, et al. *A triphasic theory for the swelling and deformation behaviors of articular cartilage.* J Biomech Eng 1991. 113:245–58. DOI: 10.1115/1.2894880 35

[238] Von der Mark, K. *Structure, biosynthesis and gene regulation of collagens in cartilage and bone,* in Dynamics of bone and cartilage metabolism, M.J. Seibel, S.P. Robins, and J.P. Bilezikian, Editors. 1999, Academic Press: San Diego. p. 3–42. 35

[239] Mendler, M, Eich-Bender, SG, et al. *Cartilage contains mixed fibrils of collagen types II, IX, and XI.* J Cell Biol 1989. 108:191–7. DOI: 10.1083/jcb.108.1.191 35

[240] Hardingham, TE. *Proteoglycans and glycosaminoglycans,* in Dynamics of bone and cartilage metabolism, M.J. Seibel, S.P. Robins, and J.P. Bilezikian, Editors. 1999, Academic Press: San Diego, CA. p. 71–81. 36, 37

[241] Maroudas, A, Bayliss, MT, et al. *Aggrecan turnover in human articular cartilage: use of aspartic acid racemization as a marker of molecular age.* Arch Biochem Biophys 1998. 350:61–71. DOI: 10.1006/abbi.1997.0492 37

[242] Verzijl, N, DeGroot, J, et al. *Age-related accumulation of the advanced glycation endproduct pentosidine in human articular cartilage aggrecan: the use of pentosidine levels as a quantitative measure of protein turnover.* Matrix Biol 2001. 20:409–17. DOI: 10.1016/S0945-053X(01)00158-5 37

[243] Woessner, JF, Jr. and Gunja-Smith, Z. *Role of metalloproteinases in human osteoarthritis.* J Rheumatol Suppl 1991. 27:99–101. 37

[244] Hudelmaier, M, Glaser, C, et al. *Age-related changes in the morphology and deformational behavior of knee joint cartilage.* Arthritis Rheum 2001. 44:2556–61. DOI: 10.1002/1529-0131(200111)44:11%3C2556::AID-ART436%3E3.0.CO;2-U 37

[245] Martin, JA and Buckwalter, JA. *Telomere erosion and senescence in human articular cartilage chondrocytes.* J Gerontol A Biol Sci Med Sci 2001. 56:B172–9. DOI: 10.1093/gerona/56.4.B172 37

[246] Buckwalter, JA, Roughley, PJ, et al. *Age-related changes in cartilage proteoglycans: quantitative electron microscopic studies.* Microsc Res Tech 1994. 28:398–408. DOI: 10.1002/jemt.1070280506 38

[247] Holmes, MW, Bayliss, MT, et al. *Hyaluronic acid in human articular cartilage. Age-related changes in content and size.* Biochem J 1988. 250:435–41. 38

[248] Buckwalter, JA and Mankin, HJ. *Articular cartilage: degeneration and osteoarthritis, repair, regeneration, and transplantation.* Instr Course Lect 1998. 47:487–504. 38

[249] Grodzinsky, AJ, Levenston, ME, et al. *Cartilage tissue remodeling in response to mechanical forces.* Annu Rev Biomed Eng 2000. 2:691–713. DOI: 10.1146/annurev.bioeng.2.1.691 38

[250] Wright, MO, Nishida, K, et al. *Hyperpolarisation of cultured human chondrocytes following cyclical pressure-induced strain: evidence of a role for alpha 5 beta 1 integrin as a chondrocyte mechanoreceptor.* J Orthop Res 1997. 15:742–7. DOI: 10.1002/jor.1100150517 38

[251] Wang, QG, Magnay, JL, et al. *Gene expression profiles of dynamically compressed single chondrocytes and chondrons.* Biochem Biophys Res Commun 2009. 379:738–42. DOI: 10.1016/j.bbrc.2008.12.111 38

[252] Lee, DA, Noguchi, T, et al. *The influence of mechanical loading on isolated chondrocytes seeded in agarose constructs.* Biorheology 2000. 37:149–61. 38

[253] Anderson, AE, Ellis, BJ, et al. *Validation of finite element predictions of cartilage contact pressure in the human hip joint.* J Biomech Eng 2008. 130:051008. DOI: 10.1115/1.2953472 38

[254] van den Bogert, AJ, Read, L, et al. *An analysis of hip joint loading during walking, running, and skiing.* Med Sci Sports Exerc 1999. 31:131–42. DOI: 10.1097/00005768-199901000-00021 38

[255] Wright, V and Dowson, D. *Lubrication and cartilage.* J Anat 1976. 121:107–18. 38, 39

[256] Swann, DA, Silver, FH, et al. *The molecular structure and lubricating activity of lubricin isolated from bovine and human synovial fluids.* Biochem J 1985. 225:195–201. 39

[257] Jay, GD, Lane, BP, et al. *Characterization of a bovine synovial fluid lubricating factor. III. The interaction with hyaluronic acid.* Connect Tissue Res 1992. 28:245–55. DOI: 10.3109/03008209209016818 39

[258] Schumacher, BL, Hughes, CE, et al. *Immunodetection and partial cDNA sequence of the proteoglycan, superficial zone protein, synthesized by cells lining synovial joints.* J Orthop Res 1999. 17:110–20. DOI: 10.1002/jor.1100170117 39

[259] Schumacher, BL, Block, JA, et al. *A novel proteoglycan synthesized and secreted by chondrocytes of the superficial zone of articular cartilage.* Arch Biochem Biophys 1994. 311:144–52. DOI: 10.1006/abbi.1994.1219 39

[260] Ogata, K, Whiteside, LA, et al. *Subchondral route for nutrition to articular cartilage in the rabbit. Measurement of diffusion with hydrogen gas in vivo.* J Bone Joint Surg Am 1978. 60:905–10. 39

[261] Maroudas, A. *Transport of solutes through cartilage: permeability to large molecules.* J Anat 1976. 122:335–47. 39

[262] Arkill, KP and Winlove, CP. *Solute transport in the deep and calcified zones of articular cartilage.* Osteoarthritis Cartilage 2008. 16:708–14. DOI: 10.1016/j.joca.2007.10.001 39

[263] Quinn, TM, Morel, V, et al. *Static compression of articular cartilage can reduce solute diffusivity and partitioning: implications for the chondrocyte biological response.* J Biomech 2001. 34:1463–9. DOI: 10.1016/S0021-9290(01)00112-9 39

[264] Evans, RC and Quinn, TM. *Solute convection in dynamically compressed cartilage.* J Biomech 2006. 39:1048–55. DOI: 10.1016/j.jbiomech.2005.02.017 39

[265] Mouritzen, U, Christgau, S, et al. *Cartilage turnover assessed with a newly developed assay measuring collagen type II degradation products: influence of age, sex, menopause, hormone replacement therapy, and body mass index.* Ann Rheum Dis 2003. 62:332–6. DOI: 10.1136/ard.62.4.332 39

[266] Fraser, A, Fearon, U, et al. *Turnover of type II collagen and aggrecan in cartilage matrix at the onset of inflammatory arthritis in humans: relationship to mediators of systemic and local inflammation.* Arthritis Rheum 2003. 48:3085–95. DOI: 10.1002/art.11331 39

[267] Mow, VC, Kuei, SC, et al. *Biphasic creep and stress relaxation of articular cartilage in compression? Theory and experiments.* J Biomech Eng 1980. 102:73–84. DOI: 10.1115/1.3138202 39

[268] Athanasiou, KA, Rosenwasser, MP, et al. *Interspecies comparisons of in situ intrinsic mechanical properties of distal femoral cartilage.* J Orthop Res 1991. 9:330–40. DOI: 10.1002/jor.1100090304 41

[269] Jurvelin, JS, Buschmann, MD, et al. *Mechanical anisotropy of the human knee articular cartilage in compression.* Proc Inst Mech Eng H 2003. 217:215–9. DOI: 10.1243/095441103765212712 41

[270] Froimson, MI, Ratcliffe, A, et al. *Differences in patellofemoral joint cartilage material properties and their significance to the etiology of cartilage surface fibrillation.* Osteoarthritis Cartilage 1997. 5:377–86. DOI: 10.1016/S1063-4584(97)80042-8 41

[271] Wayne, JS. *Load partitioning influences the mechanical response of articular cartilage.* Ann Biomed Eng 1995. 23:40–7. DOI: 10.1007/BF02368299 40

[272] Graindorge, S, Ferrandez, W, et al. *Biphasic surface amorphous layer lubrication of articular cartilage.* Med Eng Phys 2005. 27:836–44. DOI: 10.1016/j.medengphy.2005.05.001 40

[273] Ateshian, GA and Wang, H. *A theoretical solution for the frictionless rolling contact of cylindrical biphasic articular cartilage layers.* J Biomech 1995. 28:1341–55. DOI: 10.1016/0021-9290(95)00008-6 40

[274] Suh, JK. *Dynamic unconfined compression of articular cartilage under a cyclic compressive load.* Biorheology 1996. 33:289–304. DOI: 10.1016/0006-355X(96)00023-6 40

[275] Mansour, JM and Mow, VC. *The permeability of articular cartilage under compressive strain and at high pressures.* J Bone Joint Surg Am 1976. 58:509–16. 40

[276] Lai, WM and Mow, VC. *Drag-induced compression of articular cartilage during a permeation experiment.* Biorheology 1980. 17:111–23. 40

[277] Buckley, MR, Gleghorn, JP, et al. *Mapping the depth dependence of shear properties in articular cartilage.* J Biomech 2008. 41:2430–7. DOI: 10.1016/j.jbiomech.2008.05.021 42

[278] Grodzinsky, AJ. *Mechanical and electrical properties and their relevance to the physiological processes,* in Methods in cartilage research, A. Maroudas and K.E. Kuettner, Editors. 1991, Academic Press: London, England. p. 275–311. 42

[279] Mow, VC, Holmes, MH, et al. *Fluid transport and mechanical properties of articular cartilage: a review.* J Biomech 1984. 17:377–94. DOI: 10.1016/0021-9290(84)90031-9

[280] Mow, VC, Zhu, W, et al. *Structure and function of articular cartilage and meniscus*, in Basic orthopaedic biomechanics, V.C. Mow and W.C. Hayes, Editors. 1991, Raven Press: New York, NY. p. 143–198. 42

[281] Chen, MH and Broom, ND. *Concerning the ultrastructural origin of large-scale swelling in articular cartilage.* J Anat 1999. 194 (Pt 3):445–61. DOI: 10.1046/j.1469-7580.1999.19430445.x 42

[282] Maroudas, A and Venn, M. *Chemical composition and swelling of normal and osteoarthrotic femoral head cartilage. II. Swelling.* Ann Rheum Dis 1977. 36:399–406. DOI: 10.1136/ard.36.5.399 43

[283] Hollander, AP, Pidoux, I, et al. *Damage to type II collagen in aging and osteoarthritis starts at the articular surface, originates around chondrocytes, and extends into the cartilage with progressive degeneration.* J Clin Invest 1995. 96:2859–69. DOI: 10.1172/JCI118357 43

[284] Bank, RA, Soudry, M, et al. *The increased swelling and instantaneous deformation of osteoarthritic cartilage is highly correlated with collagen degradation.* Arthritis Rheum 2000. 43:2202–10. DOI: 10.1002/1529-0131(200010)43:10%3C2202::AID-ANR7%3E3.0.CO;2-E 43

[285] Roberts, S, Weightman, B, et al. *Mechanical and biochemical properties of human articular cartilage in osteoarthritic femoral heads and in autopsy specimens.* J Bone Joint Surg Br 1986. 68:278–88. 43, 78

[286] Armstrong, CG and Mow, VC. *Variations in the intrinsic mechanical properties of human articular cartilage with age, degeneration, and water content.* J Bone Joint Surg Am 1982. 64:88–94. 43

[287] Krsti*c, RV, Illustrated encyclopedia of human histology. 1984, Berlin ; New York: Springer-Verlag. viii, 450 p. 45, 48

[288] Shum, L and Nuckolls, G. *The life cycle of chondrocytes in the developing skeleton.* Arthritis Res 2002. 4:94–106. DOI: 10.1186/ar541 45, 46

[289] Hall, BK and Miyake, T. *The membranous skeleton: the role of cell condensations in vertebrate skeletogenesis.* Anat Embryol (Berl) 1992. 186:107–24. DOI: 10.1007/BF00174948 45

[290] Kim, IS, Otto, F, et al. *Regulation of chondrocyte differentiation by Cbfa1.* Mech Dev 1999. 80:159–70. DOI: 10.1016/S0925-4773(98)00210-X 45

150   REFERENCES

[291] Gerber, HP, Vu, TH, et al. *VEGF couples hypertrophic cartilage remodeling, ossification and angiogenesis during endochondral bone formation*. Nat Med 1999. 5:623–8. DOI: 10.1038/9467 46

[292] Kronenberg, HM. *Developmental regulation of the growth plate*. Nature 2003. 423:332–6. DOI: 10.1038/nature01657 46, 50

[293] Kember, NF and Sissons, HA. *Quantitative histology of the human growth plate*. J Bone Joint Surg Br 1976. 58-B:426–35. 46

[294] Hunziker, EB. *Mechanism of longitudinal bone growth and its regulation by growth plate chondrocytes*. Microsc Res Tech 1994. 28:505–19. DOI: 10.1002/jemt.1070280606 46, 48

[295] Recker, R and Barger-Lux, J. *Embryology, anatomy, and microstructure of bone*, in Disorders of bone and mineral metabolism, F.L. Coe and M.J. Favus, Editors. 2002, Lippincott Williams & Wilkins: Philadelphia. p. 177–198. 46

[296] Moore, KL, Persaud, TVN, et al., The developing human : clinically oriented embryology. 8th ed. 2008, Philadelphia, PA: Saunders/Elsevier. xiv, 522 p. 46

[297] Iannotti, J, Goldstein, S, et al. *Growth plate and bone development*, in Orthopaedic Basic Science, S.R. Simon, Editor. 1994, American Academy of Orthopaedic Surgeons: Rosemont, IL. p. 185–217. 46, 48, 50

[298] Hansson, LI, Menander-Sellman, K, et al. *Rate of normal longitudinal bone growth in the rat*. Calcif Tissue Res 1972. 10:238–51. DOI: 10.1007/BF02012553 46

[299] Brown, CC, Hembry, RM, et al. *Immunolocalization of metalloproteinases and their inhibitor in the rabbit growth plate*. J Bone Joint Surg Am 1989. 71:580–93. 46

[300] Kember, NF. *Cell population kinetics of bone growth: the first ten years of autoradiographic studies with tritiated thymidine*. Clin Orthop Relat Res 1971. 76:213–30. DOI: 10.1097/00003086-197105000-00029 48

[301] Hunziker, EB, Schenk, RK, et al. *Quantitation of chondrocyte performance in growth-plate cartilage during longitudinal bone growth*. J Bone Joint Surg Am 1987. 69:162–73. 48

[302] Hunziker, EB and Schenk, RK. *Physiological mechanisms adopted by chondrocytes in regulating longitudinal bone growth in rats*. J Physiol 1989. 414:55–71. 48

[303] Buckwalter, JA, Mower, D, et al. *Morphometric analysis of chondrocyte hypertrophy*. J Bone Joint Surg Am 1986. 68:243–55. 48

[304] Matsuzawa, T and Anderson, HC. *Phosphatases of epiphyseal cartilage studied by electron microscopic cytochemical methods*. J Histochem Cytochem 1971. 19:801–8. DOI: 10.1177/19.12.801 48

[305] Dean, DD, Muniz, OE, et al. *Localization of collagenase in the growth plate of rachitic rats.* J Clin Invest 1985. 76:716–22. DOI: 10.1172/JCI112026 48

[306] Kielty, CM, Kwan, AP, et al. *Type X collagen, a product of hypertrophic chondrocytes.* Biochem J 1985. 227:545–54. 48

[307] Mundlos, S, Engel, H, et al. *Distribution of type I and type II collagen gene expression during the development of human long bones.* Bone 1990. 11:275–9. DOI: 10.1016/8756-3282(90)90080-I 48

[308] Brighton, CT and Hunt, RM. *Histochemical localization of calcium in growth plate mitochondria and matrix vesicles.* Fed Proc 1976. 35:143–7. 48

[309] Anderson, HC. *Mechanism of mineral formation in bone.* Lab Invest 1989. 60:320–30. 48, 49

[310] Johnson, TF, Morris, DC, ct al. *Matrix vesicles and calcification of rachitic rat osteoid.* J Exp Pathol 1989. 4:123–32. 48

[311] Hohling, HJ, Steffens, H, et al. *Transmission microscopy of freeze dried, unstained epiphyseal cartilage of the guinea pig.* Cell Tissue Res 1976. 167:243–63. DOI: 10.1007/BF00224331 49

[312] Kirsch, T, Harrison, G, et al. *The roles of annexins and types II and X collagen in matrix vesicle-mediated mineralization of growth plate cartilage.* J Biol Chem 2000. 275:35577–83. DOI: 10.1074/jbc.M005648200 49

[313] Peress, NS, Anderson, HC, et al. *The lipids of matrix vesicles from bovine fetal epiphyseal cartilage.* Calcif Tissue Res 1974. 14:275–81. DOI: 10.1007/BF02060301 49

[314] Plate, U, Tkotz, T, et al. *Early mineralization of matrix vesicles in the epiphyseal growth plate.* J Microsc 1996. 183:102–7. DOI: 10.1046/j.1365-2818.1996.67430.x 49

[315] Poole, AR, Pidoux, I, et al. *Role of proteoglycans in endochondral ossification: immunofluorescent localization of link protein and proteoglycan monomer in bovine fetal epiphyseal growth plate.* J Cell Biol 1982. 92:249–60. DOI: 10.1083/jcb.92.2.249 49

[316] Buckwalter, JA and Rosenberg, L. *Structural changes in reassembled growth plate aggregates.* J Orthop Res 1986. 4:1–9. DOI: 10.1002/jor.1100040101 49

[317] MacGregor, EA and Bowness, JM. *Interaction of proteoglycans and chondroitin sulfates with calcium or phosphate ions.* Can J Biochem 1971. 49:417–25. DOI: 10.1139/o71-061 49

[318] Eanes, ED, Hailer, AW, et al. *Proteoglycan inhibition of calcium phosphate precipitation in liposomal suspensions.* Glycobiology 1992. 2:571–8. DOI: 10.1093/glycob/2.6.571 49

[319] Habuchi, H, Conrad, HE, et al. *Coordinate regulation of collagen and alkaline phosphatase levels in chick embryo chondrocytes*. J Biol Chem 1985. 260:13029–34. 49

[320] Shapiro, F, Holtrop, ME, et al. *Organization and cellular biology of the perichondrial ossification groove of ranvier: a morphological study in rabbits*. J Bone Joint Surg Am 1977. 59:703–23. 50

[321] Provot, S and Schipani, E. *Molecular mechanisms of endochondral bone development*. Biochem Biophys Res Commun 2005. 328:658–65. DOI: 10.1016/j.bbrc.2004.11.068 50

[322] Lanske, B, Karaplis, AC, et al. *PTH/PTHrP receptor in early development and Indian hedgehog-regulated bone growth*. Science 1996. 273:663–6. DOI: 10.1126/science.273.5275.663 50

[323] Kawashima-Ohya, Y, Satakeda, H, et al. *Effects of parathyroid hormone (PTH) and PTH-related peptide on expressions of matrix metalloproteinase-2, -3, and -9 in growth plate chondrocyte cultures*. Endocrinology 1998. 139:2120–7. DOI: 10.1210/en.139.4.2120 50

[324] Vu, TH, Shipley, JM, et al. *MMP-9/gelatinase B is a key regulator of growth plate angiogenesis and apoptosis of hypertrophic chondrocytes*. Cell 1998. 93:411–22. DOI: 10.1016/S0092-8674(00)81169-1 50, 59

[325] Ballock, RT, Zhou, X, et al. *Both retinoic acid and 1,25(OH)2 vitamin D3 inhibit thyroid hormone–induced terminal differentiaton of growth plate chondrocytes*. J Orthop Res 2001. 19:43–9. DOI: 10.1016/S0736-0266(00)00024-3 50

[326] Zhou, Y, Xu, BC, et al. *A mammalian model for Laron syndrome produced by targeted disruption of the mouse growth hormone receptor/binding protein gene (the Laron mouse)*. Proc Natl Acad Sci U S A 1997. 94:13215–20. DOI: 10.1073/pnas.94.24.13215 50

[327] Vortkamp, A, Lee, K, et al. *Regulation of rate of cartilage differentiation by Indian hedgehog and PTH–related protein*. Science 1996. 273:613–22. DOI: 10.1126/science.273.5275.613 50

[328] Iwasaki, M, Le, AX, et al. *Expression of indian hedgehog, bone morphogenetic protein 6 and gli during skeletal morphogenesis*. Mech Dev 1997. 69:197–202. DOI: 10.1016/S0925-4773(97)00145-7 50

[329] Minina, E, Kreschel, C, et al. *Interaction of FGF, Ihh/Pthlh, and BMP signaling integrates chondrocyte proliferation and hypertrophic differentiation*. Dev Cell 2002. 3:439–49. DOI: 10.1016/S1534-5807(02)00261-7 50

[330] Loveridge, N, Farquharson, C, et al. *Growth hormone and longitudinal bone growth in vivo: short-term effect of a growth hormone antiserum*. J Endocrinol 1995. 146:55–62. DOI: 10.1677/joe.0.1460055 51

[331] Wang, Y, Nishida, S, et al. *Insulin-like growth factor-I is essential for embryonic bone development*. Endocrinology 2006. 147:4753–61. DOI: 10.1210/en.2006-0196 51

[332] Makower, AM, Wroblewski, J, et al. *Effects of IGF-I, EGF, and FGF on proteoglycans synthesized by fractionated chondrocytes of rat rib growth plate*. Exp Cell Res 1988. 179:498–506. DOI: 10.1016/0014-4827(88)90287-X 51

[333] Ishikawa, Y, Wu, LN, et al. *Effects of calcitonin and parathyroid hormone on calcification of primary cultures of chicken growth plate chondrocytes*. J Bone Miner Res 1997. 12:356–66. DOI: 10.1359/jbmr.1997.12.3.356 51

[334] Baron, J, Huang, Z, et al. *Dexamethasone acts locally to inhibit longitudinal bone growth in rabbits*. Am J Physiol 1992. 263:E489–92. 51

[335] Silvestrini, G, Ballanti, P, et al. *Evaluation of apoptosis and the glucocorticoid receptor in the cartilage growth plate and metaphyseal bone cells of rats after high-dose treatment with corticosterone*. Bone 2000. 26:33–42. DOI: 10.1016/S8756-3282(99)00245-8 51

[336] Boyan, BD, Schwartz, Z, et al. *In vitro studies on the regulation of endochondral ossification by vitamin D*. Crit Rev Oral Biol Med 1992. 3:15–30. 51

[337] Swain, LD, Schwartz, Z, et al. *Nongenomic regulation of chondrocyte membrane fluidity by 1,25-(OH)2D3 and 24,25-(OH)2D3 is dependent on cell maturation*. Bone 1993. 14:609–17. DOI: 10.1016/8756-3282(93)90082-L 51

[338] Fernandez-Cancio, M, Andaluz, P, et al. *Vitamin D stimulates growth hormone–insulin–like growth factor (GH-IGF) gene axis expression and potentiates GH effect to reverse the inhibition produced by glucocorticoids in human growth plate chondrocytes*. Hormone Research 2007. 67:204–205 DOI: 10.1159/000097582 51

[339] De Luca, F, Uyeda, JA, et al. *Retinoic acid is a potent regulator of growth plate chondrogenesis*. Endocrinology 2000. 141:346–53. DOI: 10.1210/en.141.1.346 51

[340] Steele, RG, Lugg, P, et al. *Premature epiphyseal closure secondary to single-course vitamin A therapy*. Aust N Z J Surg 1999. 69:825–7. DOI: 10.1046/j.1440-1622.1999.01706.x 51

[341] Hill, DJ, Logan, A, et al. *Stimulation of DNA and protein synthesis in epiphyseal growth plate chondrocytes by fibroblast growth factors. Interactions with other peptide growth factors*. Ann N Y Acad Sci 1991. 638:449–52. DOI: 10.1111/j.1749-6632.1991.tb49066.x 51

[342] Wroblewski, J and Edwall, C. *PDGF BB stimulates proliferation and differentiation in cultured chondrocytes from rat rib growth plate*. Cell Biol Int Rep 1992. 16:133–44. DOI: 10.1016/S0309-1651(06)80107-4 51

[343] Choi, YC, Morris, GM, et al. *Effect of platelet lysate on growth and sulfated glycosaminoglycan synthesis in articular chondrocyte cultures*. Arthritis Rheum 1980. 23:220–4. DOI: 10.1002/art.1780230213 51

154 REFERENCES

[344] O'Rahilly, R and Gardner, E. *The timing and sequence of events in the development of the limbs in the human embryo.* Anat Embryol (Berl) 1975. 148:1–23. DOI: 10.1007/BF00315559 51

[345] Ogden, JA. *Chondro-osseus development and growth*, in Fundamental and clinical bone physiology, M.R. Urist, Editor. 1980, J.B. Lippincott Company: Philadelphia, PA. p. 108–171. 52

[346] Onyekwelu, I, Goldring, MB, et al. *Chondrogenesis, joint formation, and articular cartilage regeneration.* J Cell Biochem 2009. 107:383–92. DOI: 10.1002/jcb.22149 52, 78

[347] Schaffler, MB. *Role of bone turnover in microdamage.* Osteoporos Int 2003. 14 Suppl 5:S73–7; discussion S77–80. DOI: 10.1007/s00198-003-1477-1 53

[348] Frost, HH. *Presence of microscopic cracks in vivo in bone.* Henry Ford Medical Bulletin 1960. 8:27–35. 53

[349] Mashiba, T, Turner, CH, et al. *Effects of suppressed bone turnover by bisphosphonates on microdamage accumulation and biomechanical properties in clinically relevant skeletal sites in beagles.* Bone 2001. 28:524–31. DOI: 10.1016/S8756-3282(01)00414-8 53

[350] Burr, DB. *Targeted and nontargeted remodeling.* Bone 2002. 30:2–4. DOI: 10.1016/S8756-3282(01)00619-6 53

[351] Verborgt, O, Gibson, GJ, et al. *Loss of osteocyte integrity in association with microdamage and bone remodeling after fatigue in vivo.* J Bone Miner Res 2000. 15:60–7. DOI: 10.1359/jbmr.2000.15.1.60 53

[352] Cardoso, L, Herman, BC, et al. *Osteocyte apoptosis controls activation of intracortical resorption in response to bone fatigue.* J Bone Miner Res 2009. 24:597–605. DOI: 10.1359/jbmr.081210 53

[353] Nordin, M and Frankel, VH, Basic biomechanics of the musculoskeletal system. 3rd ed. 2001, Philadelphia: Lippincott Williams & Wilkins. xvii, 467 p. 54

[354] Simon, RR, Sherman, SC, et al., Emergency orthopedics : the extremities. 5th ed. 2007, New York: McGraw-Hill, Medical Pub. Division. xxxvi, 556 p. 54

[355] Perren, SM, Huggler, A, et al. *The reaction of cortical bone to compression.* Acta Orthop Scand Suppl 1969. 125:19–29. 55

[356] Heppenstall, RB. *Fracture Healing*, in Soft and hard tissue repair : biological and clinical aspects, T.K. Hunt, Editor. 1984, Praeger: New York. p. 101–142. 55

[357] Tonna, EA and Pentel, L. *Chondrogenic cell formation via osteogenic cell progeny transformation.* Lab Invest 1972. 27:418–26. 55

[358] Wray, JB. *Vascular regeneration in the healing fracture. An experimental study.* Angiology 1963. 14:134–8. DOI: 10.1177/000331976301400306 55

[359] Urist, MR, Wallace, TH, et al. *The Function of Fibrocartilaginous Fracture Callus. Observations on Transplants Labelled with Tritiated Thymidine.* J Bone Joint Surg Br 1965. 47:304–18. 55

[360] Sumner-Smith, G, Bone in clinical orthopaedics : a study in comparative osteology. 1982, Philadelphia: Saunders. xvi, 435 p. 56

[361] Lu, C, Hansen, E, et al. *Effect of age on vascularization during fracture repair.* J Orthop Res 2008. 26:1384–9. DOI: 10.1002/jor.20667 56

[362] Ow, A and Cheung, LK. *Skeletal stability and complications of bilateral sagittal split osteotomies and mandibular distraction osteogenesis: an evidence-based review.* J Oral Maxillofac Surg 2009. 67:2344–53. DOI: 10.1016/j.joms.2008.07.003 56

[363] Neovius, E and Engstrand, T. *Craniofacial reconstruction with bone and biomaterials: review over the last 11 years.* J Plast Reconstr Aesthet Surg 2010. 63:1615–23. DOI: 10.1016/j.bjps.2009.06.003 56

[364] Saulacic, N, Zix, J, et al. *Complication rates and associated factors in alveolar distraction osteogenesis: a comprehensive review.* Int J Oral Maxillofac Surg 2009. 38:210–7. DOI: 10.1016/j.ijom.2009.01.002 56

[365] Codivilla, A. *On the means of lengthening, in the lower limbs, the muscles and tissues which are shortened through deformity. 1904.* Clin Orthop Relat Res 1994. 4–9. DOI: 10.1007/s11999-008-0518-7 56

[366] Ilizarov, GA. *The tension-stress effect on the genesis and growth of tissues. Part I. The influence of stability of fixation and soft-tissue preservation.* Clin Orthop Relat Res 1989. 249–81. 56

[367] Ilizarov, GA. *The tension-stress effect on the genesis and growth of tissues: Part II. The influence of the rate and frequency of distraction.* Clin Orthop Relat Res 1989. 263–85. 56

[368] Tay, BK, Le, AX, et al. *Histochemical and molecular analyses of distraction osteogenesis in a mouse model.* J Orthop Res 1998. 16:636–42. DOI: 10.1002/jor.1100160518 56

[369] Glowacki, J, Schulten, AJ, et al. *Nicotine impairs distraction osteogenesis in the rat mandible.* Int J Oral Maxillofac Surg 2008. 37:156–61. DOI: 10.1016/j.ijom.2007.08.001 56, 92

[370] Zheng, LW, Ma, L, et al. *Changes in blood perfusion and bone healing induced by nicotine during distraction osteogenesis.* Bone 2008. 43:355–61. DOI: 10.1016/j.bone.2008.04.002 56

[371] Aronson, J and Shen, X. *Experimental healing of distraction osteogenesis comparing metaphyseal with diaphyseal sites.* Clin Orthop Relat Res 1994. 25–30. 56

[372] Glowacki, J, Shusterman, EM, et al. *Distraction osteogenesis of the porcine mandible: histomorphometric evaluation of bone.* Plast Reconstr Surg 2004. 113:566–73. DOI: 10.1097/01.PRS.0000101061.99577.09 56, 88, 92

[373] Hall, BK. *Earliest evidence of cartilage and bone development in embryonic life.* Clin Orthop Relat Res 1987. 255–72. 57

[374] Karsenty, G. *Transcriptional control of skeletogenesis.* Annu Rev Genomics Hum Genet 2008. 9:183–96. DOI: 10.1146/annurev.genom.9.081307.164437 58

[375] Karaplis, AC, Luz, A, et al. *Lethal skeletal dysplasia from targeted disruption of the parathyroid hormone-related peptide gene.* Genes Dev 1994. 8:277–89. DOI: 10.1101/gad.8.3.277 58

[376] Hartmann, C. *A Wnt canon orchestrating osteoblastogenesis.* Trends Cell Biol 2006. 16:151–8. DOI: 10.1016/j.tcb.2006.01.001 58

[377] Li, X and Cao, X. *BMP signaling and skeletogenesis.* Ann N Y Acad Sci 2006. 1068:26–40. DOI: 10.1196/annals.1346.006 58

[378] Ashton, IK and Dekel, S. *Fracture repair in the Snell dwarf mouse.* Br J Exp Pathol 1983. 64:479–86. 59

[379] Gerstenfeld, LC, Cho, TJ, et al. *Impaired fracture healing in the absence of TNF-alpha signaling: the role of TNF-alpha in endochondral cartilage resorption.* J Bone Miner Res 2003. 18:1584–92. DOI: 10.1359/jbmr.2003.18.9.1584 59

[380] Colnot, C, Thompson, Z, et al. *Altered fracture repair in the absence of MMP9.* Development 2003. 130:4123–33. DOI: 10.1242/dev.00559 59

[381] Pasparakis, M, Alexopoulou, L, et al. *Immune and inflammatory responses in TNF alpha-deficient mice: a critical requirement for TNF alpha in the formation of primary B cell follicles, follicular dendritic cell networks and germinal centers, and in the maturation of the humoral immune response.* J Exp Med 1996. 184:1397–411. DOI: 10.1084/jem.184.4.1397 59

[382] Peschon, JJ, Torrance, DS, et al. *TNF receptor-deficient mice reveal divergent roles for p55 and p75 in several models of inflammation.* J Immunol 1998. 160:943–52. 59

[383] Bandyopadhyay, A, Tsuji, K, et al. *Genetic analysis of the roles of BMP2, BMP4, and BMP7 in limb patterning and skeletogenesis.* PLoS Genet 2006. 2:e216. DOI: 10.1371/journal.pgen.0020216 59

[384] Tsuji, K, Bandyopadhyay, A, et al. *BMP2 activity, although dispensable for bone formation, is required for the initiation of fracture healing.* Nat Genet 2006. 38:1424–9. DOI: 10.1038/ng1916 59

[385] Olutoye, OO and Cohen, IK. *Fetal wound healing: an overview.* Wound Repair Regen 1996. 4:66–74. DOI: 10.1046/j.1524-475X.1996.40112.x 60

[386] Colwell, AS, Krummel, TM, et al. *An in vivo mouse excisional wound model of scarless healing.* Plast Reconstr Surg 2006. 117:2292–6. DOI: 10.1097/01.prs.0000219340.47232.eb 60

[387] Longaker, MT, Whitby, DJ, et al. *Fetal diaphragmatic wounds heal with scar formation.* J Surg Res 1991. 50:375–85. DOI: 10.1016/0022-4804(91)90206-2 60

[388] Buchanan, EP, Longaker, MT, et al. *Fetal skin wound healing.* Adv Clin Chem 2009. 48:137–61. DOI: 10.1016/S0065-2423(09)48006-5 60

[389] Longaker, MT, Adzick, NS, et al. *Studies in fetal wound healing, VII. Fetal wound healing may be modulated by hyaluronic acid stimulating activity in amniotic fluid.* J Pediatr Surg 1990. 25:430–3. DOI: 10.1016/0022-3468(90)90387-O 60

[390] Shah, M, Foreman, DM, et al. *Neutralisation of TGF-beta 1 and TGF-beta 2 or exogenous addition of TGF-beta 3 to cutaneous rat wounds reduces scarring.* J Cell Sci 1995. 108 (Pt 3):985–1002. 60

[391] Rhett, JM, Ghatnekar, GS, et al. *Novel therapies for scar reduction and regenerative healing of skin wounds.* Trends Biotechnol 2008. 26:173–80. DOI: 10.1016/j.tibtech.2007.12.007 60

[392] Longaker, MT, Moelleken, BR, et al. *Fetal fracture healing in a lamb model.* Plast Reconstr Surg 1992. 90:161–71; discussion 172-3. DOI: 10.1097/00006534-199290020-00001 60

[393] Glowacki, J, Lackey, M, et al. Differences in fetal and post-natal bone repair and responses to implants. . in Second International Conference on Bone Morphogenetic Proteins. 1997. Sacramento, CA. 60, 86, 89

[394] Eid, K, Zelicof, S, et al. *Tissue reactions to particles of bone-substitute materials in intraosseous and heterotopic sites in rats: discrimination of osteoinduction, osteocompatibility, and inflammation.* J Orthop Res 2001. 19:962–9. DOI: 10.1016/S0736-0266(00)00080-2 61, 63, 85, 99

[395] *The burden of musculoskeletal diseases in the United States.* Published by: Bone and Joint Decade. Last Updated: 2010. Accessed on: November 17, 2010. Available at: http://www.boneandjointburden.org/. 61, 76

[396] Audige, L, Griffin, D, et al. *Path analysis of factors for delayed healing and nonunion in 416 operatively treated tibial shaft fractures.* Clin Orthop Relat Res 2005. 438:221–32. DOI: 10.1097/01.blo.0000163836.66906.74 61

[397] Burchardt, H. *Biology of bone transplantation.* Orthop Clin North Am 1987. 18:187–96. 62

158 REFERENCES

[398] Chen, NT, Glowacki, J, et al. *The roles of revascularization and resorption on endurance of craniofacial onlay bone grafts in the rabbit.* Plast Reconstr Surg 1994. 93:714–22; discussion 723–4. DOI: 10.1097/00006534-199404000-00009 62

[399] Glowacki, J, Kaban, LB, et al. *Physiological aspects of bone repair using demineralized bone*, in Soft and hard tissue repair : biological and clinical aspects, T.K. Hunt, Editor. 1984, Praeger: New York. p. 265–280. 62

[400] Glowacki, J and Mulliken, JB. *Demineralized bone implants.* Clin Plast Surg 1985. 12:233–41. 62

[401] De Long, WG, Jr., Einhorn, TA, et al. *Bone grafts and bone graft substitutes in orthopaedic trauma surgery. A critical analysis.* J Bone Joint Surg Am 2007. 89:649–58. DOI: 10.2106/JBJS.F.00465 63

[402] Glowacki, J. *A review of osteoinductive testing methods and sterilization processes for demineralized bone.* Cell Tissue Bank 2005. 6:3–12. DOI: 10.1007/s10561-005-4252-z 63, 91

[403] Goldberg, VM. *Biology of bone allograft and clinical application*, in Musculoskeletal tissue regeneration : biological materials and methods, W.S. Pietrzak, Editor. 2008, Humana Press: Totowa, NJ. p. 81–92. 63

[404] Urist, MR. *Bone: formation by autoinduction.* Science 1965. 150:893–9. DOI: 10.1126/science.150.3698.893 63, 90, 102

[405] Reddi, AH and Huggins, C. *Biochemical sequences in the transformation of normal fibroblasts in adolescent rats.* Proc Natl Acad Sci U S A 1972. 69:1601–5. DOI: 10.1073/pnas.69.6.1601 63

[406] Glowacki, J, Kaban, LB, et al. *Application of the biological principle of induced osteogenesis for craniofacial defects.* Lancet 1981. 1:959–62. DOI: 10.1016/S0140-6736(81)91730-X 63

[407] Kaban, LB, Mulliken, JB, et al. *Treatment of jaw defects with demineralized bone implants.* J Oral Maxillofac Surg 1982. 40:623–6. DOI: 10.1016/0278-2391(82)90109-4 63

[408] Sonis, ST, Kaban, LB, et al. *Clinical trial of demineralized bone powder in the treatment of periodontal defects.* J Oral Med 1983. 38:117–22. 63

[409] Upton, J and Glowacki, J. *Hand reconstruction with allograft demineralized bone: twenty-six implants in twelve patients.* J Hand Surg Am 1992. 17:704–13. DOI: 10.1016/0363-5023(92)90321-F 63

[410] Rosenthal, RK, Folkman, J, et al. *Demineralized bone implants for nonunion fractures, bone cysts, and fibrous lesions.* Clin Orthop Relat Res 1999. 61–9. DOI: 10.1097/00003086-199907000-00009 63

[411] Hench, L and Best, S. *Ceramics, Glasses and Glass-Ceramics*, in Biomaterials science : an introduction to materials in medicine, B.D. Ratner, Editor. 2004, Elsevier Academic Press: Amsterdam ; Boston. p. 153–170. 63

[412] Pettis, GY, Kaban, LB, et al. *Tissue response to composite ceramic hydroxyapatite/demineralized bone implants*. J Oral Maxillofac Surg 1990. 48:1068–74. DOI: 10.1016/0278-2391(90)90291-9 63, 90, 92

[413] Khan, Y, Yaszemski, MJ, et al. *Tissue engineering of bone: material and matrix considerations*. J Bone Joint Surg Am 2008. 90 Suppl 1:36–42. DOI: 10.2106/JBJS.G.01260 63, 97, 98, 99, 100

[414] Chang, CC and Merritt, K. *Infection at the site of implanted materials with and without preadhered bacteria*. J Orthop Res 1994. 12:526–31. DOI: 10.1002/jor.1100120409 63

[415] Rosengren, A, Wallman, L, et al. *Tissue reactions evoked by porous and plane surfaces made out of silicon and titanium*. IEEE Trans Biomed Eng 2002. 49:392–9. DOI: 10.1109/10.991167 63

[416] Tang, L, Jennings, TA, et al. *Mast cells mediate acute inflammatory responses to implanted biomaterials*. Proc Natl Acad Sci U S A 1998. 95:8841–6. DOI: 10.1073/pnas.95.15.8841 63

[417] Santavirta, S, Gristina, A, et al. *Cemented versus cementless hip arthroplasty. A review of prosthetic biocompatibility*. Acta Orthop Scand 1992. 63:225–32. DOI: 10.3109/17453679209154831 63

[418] Anderson, JM. *Inflammatory response to implants*. ASAIO Trans 1988. 34:101–7. DOI: 10.1097/00002480-198804000-00005 63

[419] Nuss, KM and von Rechenberg, B. *Biocompatibility issues with modern implants in bone - a review for clinical orthopedics*. Open Orthop J 2008. 2:66–78. DOI: 10.2174/1874325000802010066 64

[420] Matukas, VJ. *Medical risks associated with dental implants*. J Dent Educ 1988. 52:745–7. 64

[421] Adell, R. *The surgical principles of osseointegration*, in Advanced osseointegration surgery : applications in the maxillofacial region, P. Worthington and P.-I. Brånemark, Editors. 1992, Quintessence Books: Chicago. p. 94–107. 64

[422] Esposito, M, Hirsch, JM, et al. *Biological factors contributing to failures of osseointegrated oral implants. (II). Etiopathogenesis*. Eur J Oral Sci 1998. 106:721–64. DOI: 10.1046/j.0909-8836..t01-6-.x 64

160   REFERENCES

[423] Goldring, SR, Schiller, AL, et al. *The synovial-like membrane at the bone-cement interface in loose total hip replacements and its proposed role in bone lysis.* J Bone Joint Surg Am 1983. 65:575–84. 64

[424] Campbell, P, Ma, S, et al. *Isolation of predominantly submicron-sized UHMWPE wear particles from periprosthetic tissues.* J Biomed Mater Res 1995. 29:127–31. DOI: 10.1002/jbm.820290118 64

[425] Glowacki, J, Jasty, M, et al. *Comparison of multinucleated cells elicited in rats by particulate bone, polyethylene, or polymethylmethacrylate.* J Bone Miner Res 1986. 1:327–31. DOI: 10.1002/jbmr.5650010405 64

[426] Gray, MH, Talbert, ML, et al. *Changes seen in lymph nodes draining the sites of large joint prostheses.* Am J Surg Pathol 1989. 13:1050–6. DOI: 10.1097/00000478-198912000-00007 64

[427] McNie, CM, Barton, DC, et al. *The prediction of polyethylene wear rate and debris morphology produced by microscopic asperities on femoral heads.* J Mater Sci Mater Med 2000. 11:163–74. DOI: 10.1023/A:1008928219843 64

[428] Athanasiou, KA, Darling, E, et al., Articular cartilage tissue engineering, in Synthesis Lectures on Tissue Engineering, K.A. Athanasiou, Editor. 2009, Morgan & Claypool: San Rafael, CA. DOI: 10.2200/S00212ED1V01Y200910TIS003 64

[429] Kreder, HJ, Moran, M, et al. *Biologic resurfacing of a major joint defect with cryopreserved allogeneic periosteum under the influence of continuous passive motion in a rabbit model.* Clin Orthop Relat Res 1994. 288–96. 64

[430] Convery, FR, Akeson, WH, et al. *The repair of large osteochondral defects. An experimental study in horses.* Clin Orthop Relat Res 1972. 82:253–62. DOI: 10.1097/00003086-197201000-00033 64

[431] Behrens, F, Shepard, N, et al. *Metabolic recovery of articular cartilage after intra-articular injections of glucocorticoid.* J Bone Joint Surg Am 1976. 58:1157–60. 65

[432] Buckwalter, JA, Wossenbery, L, et al. *Articular cartilage injury and repair,* in Injury and repair of the musculoskeletal soft tissues S.L.Y. Woo and J.A. Buckwalter, Editors. 1988, American Academy of Orthopaedic Surgeons: Park Ridge, Ill. p. 465–482. 65

[433] Reddi, AH. *Extracellular bone matrix dependent local induction of cartilage and bone.* J Rheumatol Suppl 1983. 11:67–9. 65

[434] Weinstein, SL, Buckwalter, JA, et al., Turek's Orthopaedics : principles and their application. 5th ed. 1994, Philadelphia: Lippincott. xiv, 708 p., 8 p. of plates. 65

[435] Nehrer, S, Spector, M, et al. *Histologic analysis of tissue after failed cartilage repair procedures.* Clin Orthop Relat Res 1999. 149–62. DOI: 10.1097/00003086-199908000-00020 66

[436] Gardner, E and O'Rahilly, R. *The early development of the knee joint in staged human embryos.* J Anat 1968. 102:289–99. 66

[437] *Craniofacial Anomalies.* Published by: Children's Hospital Boston. Last Updated: 2010. Accessed on: November 2010. Available at: http://www.childrenshospital.org/az/Site1407/mainpageS1407P0.html. 67

[438] Kaban, LB, Pediatric oral and maxillofacial surgery. 1990, Philadelphia: Saunders. xi, 493 p. 67

[439] Kaban, LB, Padwa, BL, et al. *Surgical correction of mandibular hypoplasia in hemifacial microsomia: the case for treatment in early childhood.* J Oral Maxillofac Surg 1998. 56:628–38. DOI: 10.1016/S0278-2391(98)90465-7 68

[440] Kaban, LB, Moses, MH, et al. *Surgical correction of hemifacial microsomia in the growing child.* Plast Reconstr Surg 1988. 82:9–19. DOI: 10.1097/00006534-198882010-00003 68

[441] Mulliken, JB, Kaban, LB, et al. *Facial skeletal changes following hypertelorbitism correction.* Plast Reconstr Surg 1986. 77:7–16. DOI: 10.1097/00006534-198601000-00003 68

[442] *Developmental dysplasia of the hip.* Published by: MedLine Plus. Last Updated: 2010. Accessed on: November 2010. Available at: http://www.nlm.nih.gov/medlineplus/ency/article/000971.htm. 68

[443] *Simple Bone Cyst.* Published by: Children's Hospital Boston. Last Updated: 2010. Accessed on: November 2010. Available at: http://www.childrenshospital.org/az/Site642/mainpageS642P0.html. 68

[444] Wright, JG, Yandow, S, et al. *A randomized clinical trial comparing intralesional bone marrow and steroid injections for simple bone cysts.* J Bone Joint Surg Am 2008. 90:722–30. DOI: 10.2106/JBJS.G.00620 69

[445] Eastwood, B (2010) *Aneurysmal Bone Cyst.* eMedicine. Available at: http://emedicine.medscape.com/article/1254784-overview. 69

[446] Lewis, VO. *What's new in musculoskeletal oncology.* J Bone Joint Surg Am 2009. 91:1546–56. DOI: 10.2106/JBJS.I.00375 69, 70, 71

[447] Lin, PP, Brown, C, et al. *Aneurysmal bone cysts recur at juxtaphyseal locations in skeletally immature patients.* Clin Orthop Relat Res 2008. 466:722–8. DOI: 10.1007/s11999-007-0080-8 69

162   REFERENCES

[448] *Giant Cell Tumor*. Published by: Children's Hospital Boston. Last Updated: 2010. Accessed on: November 2010. Available at: http://www.childrenshospital.org/az/Site960/ mainpageS960P0.html. 69

[449] *Osteoid Osteoma*. Published by: Children's Hospital Boston. Last Updated: 2010. Accessed on: November 2010. Available at: http://www.childrenshospital.org/az/Site1078/ mainpageS1078P0.html. 69

[450] *Chondromas*. Published by: Children's Hospital Boston. Last Updated: 2010. Accessed on: November 2010. Available at: http://www.childrenshospital.org/az/Site746/ mainpageS746P0.html. 69

[451] *Adamantioma*. Published by: Children's Hospital Boston. Last Updated: 2010. Accessed on: November 2010. Available at: http://www.childrenshospital.org/az/Site760/ mainpageS760P0.html. 69

[452] *Information about bone cancer: chondrosarcoma*. Published by: Stanford Medicine. Last Updated: 2010. Accessed on: November 2010. Available at: http://cancer.stanford.edu/ bonecancer/bone/chondro.html. 70

[453] *Osteosarcoma*. Published by: Children's Hospital Boston. Last Updated: 2010. Accessed on: November 2010. Available at: http://www.childrenshospital.org/az/Site1077/ mainpageS1077P0.html. 70

[454] *Ewing's Sarcoma*. Published by: Children's Hospital Boston. Last Updated: 2010. Accessed on: November 2010. Available at: http://www.childrenshospital.org/az/Site852/ mainpageS852P0.html. 70

[455] Barrett, D, Fish, JD, et al. *Autologous and allogeneic cellular therapies for high-risk pediatric solid tumors*. Pediatr Clin North Am 2010. 57:47–66. DOI: 10.1016/j.pcl.2010.01.001 70

[456] Balogh, JM and Sutherland, SE. *Osteoradionecrosis of the mandible: a review*. J Otolaryngol 1989. 18:245–50. 70

[457] Epstein, JB, Wong, FL, et al. *Osteoradionecrosis: clinical experience and a proposal for classification*. J Oral Maxillofac Surg 1987. 45:104–10. DOI: 10.1016/0278-2391(87)90399-5 70

[458] Blanchaert, RH and Harris, CM (2010) *Osteoradionecrosis of the mandible*. eMedicine. Available at: http://emedicine.medscape.com/article/851539-overview. 70

[459] Freiberger, JJ and Feldmeier, JJ. *Evidence supporting the use of hyperbaric oxygen in the treatment of osteoradionecrosis of the jaw*. J Oral Maxillofac Surg 2010. 68:1903–6. DOI: 10.1016/j.joms.2010.02.001 70

[460] Bessereau, J and Annane, D. *Treatment of osteoradionecrosis of the jaw: the case against the use of hyperbaric oxygen.* J Oral Maxillofac Surg 2010. 68:1907–10.
DOI: 10.1016/j.joms.2010.02.005 70

[461] Cannon, CP, Lin, PP, et al. *Management of radiation-associated fractures.* J Am Acad Orthop Surg 2008. 16:541–9. 70

[462] Ferguson, HW and Stevens, MR. *Advances in head and neck radiotherapy to the mandible.* Oral Maxillofac Surg Clin North Am 2007. 19:553–63, vii. DOI: 10.1016/j.coms.2007.07.005 71

[463] Greenberger, JS and Epperly, MW. *Review. Antioxidant gene therapeutic approaches to normal tissue radioprotection and tumor radiosensitization.* In Vivo 2007. 21:141–6. 71

[464] Weldon, D. *The effects of corticosteroids on bone: osteonecrosis (avascular necrosis of the bone).* Ann Allergy Asthma Immunol 2009. 103:91–7; quiz 97-100, 133.
DOI: 10.1016/S1081-1206(10)60159-7 71

[465] Boss, JH and Misselevich, I. *Osteonecrosis of the femoral head of laboratory animals: the lessons learned from a comparative study of osteonecrosis in man and experimental animals.* Vet Pathol 2003. 40:345–54. DOI: 10.1354/vp.40-4-345 71

[466] Filleul, O, Crompot, E, et al. *Bisphosphonate-induced osteonecrosis of the jaw: a review of 2,400 patient cases.* J Cancer Res Clin Oncol 2010. 136:1117–24.
DOI: 10.1007/s00432-010-0907-7 71

[467] Goh, SK, Yang, KY, et al. *Subtrochanteric insufficiency fractures in patients on alendronate therapy: a caution.* J Bone Joint Surg Br 2007. 89:349–53.
DOI: 10.1302/0301-620X.89B3.18146 71

[468] *Osteogenesis Imperfecta* Published by: Osteogenesis Imperfecta Foundation. Last Updated: 2007. Accessed on: November 2010. Available at: http://www.oif.org. 71

[469] Basel, D and Steiner, RD. *Osteogenesis imperfecta: recent findings shed new light on this once well-understood condition.* Genet Med 2009. 11:375–85. DOI: 10.1097/GIM.0b013e3181a1ff7b 71

[470] Glorieux, FH. *Experience with bisphosphonates in osteogenesis imperfecta.* Pediatrics 2007. 119 Suppl 2:S163–5. DOI: 10.1542/peds.2006-2023I 72

[471] Kamoun-Goldrat, AS and Le Merrer, MF. *Animal models of osteogenesis imperfecta and related syndromes.* J Bone Miner Metab 2007. 25:211–8. DOI: 10.1007/s00774-007-0750-3 72

[472] Millington-Ward, S, McMahon, HP, et al. *Emerging therapeutic approaches for osteogenesis imperfecta.* Trends Mol Med 2005. 11:299–305. DOI: 10.1016/j.molmed.2005.04.006 72

[473] Bhan, A, Rao, AD, et al. *Osteomalacia as a result of vitamin D deficiency*. Endocrinol Metab Clin North Am 2010. 39:321–31, table of contents. DOI: 10.1016/j.ecl.2010.02.001 72

[474] DeLucia, MC, Mitnick, ME, et al. *Nutritional rickets with normal circulating 25-hydroxyvitamin D: a call for reexamining the role of dietary calcium intake in North American infants*. J Clin Endocrinol Metab 2003. 88:3539–45. DOI: 10.1210/jc.2002-021935 72

[475] Greer, FR and Finberg, L (2010) *Rickets*. eMedicine. Available at: http://emedicine. medscape.com/article/985510-overview. 72

[476] de Torrente de la Jara, G, Pecoud, A, et al. *Musculoskeletal pain in female asylum seekers and hypovitaminosis D3*. BMJ 2004. 329:156–7. DOI: 10.1136/bmj.329.7458.156 72

[477] LeBoff, MS, Kohlmeier, L, et al. *Occult vitamin D deficiency in postmenopausal US women with acute hip fracture*. JAMA 1999. 281:1505–11. DOI: 10.1001/jama.281.16.1505 72

[478] Carpenter, TO. *Oncogenic osteomalacia–a complex dance of factors*. N Engl J Med 2003. 348:1705–8. DOI: 10.1056/NEJMe030037 73

[479] Salen, PN (2010) *Hyperparathyroidism*. eMedicine. Available at: http://emedicine. medscape.com/article/766906-overview. 73

[480] *National Osteporosis Foundation Home Page*. Published by. Last Updated: 2010. Accessed on: november 2010. Available at: http://www.nof.org. 73

[481] Kanis, JA, Melton, LJ, 3rd, et al. *The diagnosis of osteoporosis*. J Bone Miner Res 1994. 9:1137–41. DOI: 10.1002/jbmr.5650090802 73

[482] Glowacki, J. *Osteoarthritis and osteoporosis: Coexistence of osteoporosis in patients with osteoarthritis*. Minerva Orthop Traum 2010. 61:115–122. 73, 79

[483] Bouxsein, ML and Seeman, E. *Quantifying the material and structural determinants of bone strength*. Best Pract Res Clin Rheumatol 2009. 23:741–53. DOI: 10.1016/j.berh.2009.09.008 73

[484] *Managing & Treating Osteoporosis. Osteoporosis Medicines: What You Need to Know*. Published by. Last Updated: 2010. Accessed on: November 2010. Available 74

[485] Kream, BE, Graves, L, et al. *Clinical and basic aspects of glucocorticoid action in bone*, in Principles of bone biology, J.P. Bilezikian, L.G. Raisz, and T.J. Martin, Editors. 2008, Academic Press/Elsevier: San Diego, CA. p. 955–981. 74

[486] *Glucocorticoid-Induced Osteoporosis*. Published by: American College of Rheumatology. Last Updated: 2010. Accessed on: November 2010. Available at: http://www. rheumatology.org/practice/clinical/patients/diseases_and_conditions/ gi-osteoporosis.asp. 74

[487] Olney, RC. *Mechanisms of impaired growth: effect of steroids on bone and cartilage*. Horm Res 2009. 72 Suppl 1:30–5. DOI: 10.1159/000229761 74

[488] Bassett, JH and Williams, GR. *The molecular actions of thyroid hormone in bone*. Trends Endocrinol Metab 2003. 14:356–64. DOI: 10.1016/S1043-2760(03)00144-9 74

[489] Serota, AC and Lane, JM (2010) *Osteoporosis (secondary)*. eMedicine. Available at: http://emedicine.medscape.com/article/311449-overview. 74

[490] *What people with diabetes need to know about osteoporosis*. Published by: NIH Osteoporosis and related bone diseases national resource center. Last Updated: 2010. Accessed on: November 2010. Available at: http://www.niams.nih.gov/Health_Info/Bone/Osteoporosis/Conditions_Behaviors/diabetes.asp. 75

[491] Schneider, S (2002) *Diabetes and bone mass*. Medscape Diabetes and Endocrinology. Available at: http://www.medscape.com/viewarticle/433252. 75

[492] Gaston, MS and Simpson, AI I. *Inhibition of fracture healing*. J Bone Joint Surg Br 2007. 89:1553–60. DOI: 10.1302/0301-620X.89B12.19671 75

[493] Kasahara, T, Imai, S, et al. *Malfunction of bone marrow–derived osteoclasts and the delay of bone fracture healing in diabetic mice*. Bone 2010. 47:617–25. DOI: 10.1016/j.bone.2010.06.014 75

[494] Verhaeghe, J and Bouillon, R. *Effects of diabetes and insulin on bone physiology*, in Principles of bone biology, J.P. Bilezikian, L.G. Raisz, and T.J. Martin, Editors. 2008, Academic Press/Elsevier: San Diego, CA. p. 983–999. 75

[495] Lucas, PD. *Reversible reduction in bone blood flow in streptozotocin-diabetic rats*. Experientia 1987. 43:894–5. DOI: 10.1007/BF01951655 75

[496] Santana, RB, Xu, L, et al. *A role for advanced glycation end products in diminished bone healing in type 1 diabetes*. Diabetes 2003. 52:1502–10. DOI: 10.2337/diabetes.52.6.1502 75

[497] Habib, ZA, Havstad, SL, et al. *Thiazolidinedione use and the longitudinal risk of fractures in patients with type 2 diabetes mellitus*. J Clin Endocrinol Metab 2010. 95:592–600. DOI: 10.1210/jc.2009-1385 75

[498] Grey, A. *Thiazolidinedione-induced skeletal fragility–mechanisms and implications*. Diabetes Obes Metab 2009. 11:275–84. DOI: 10.1111/j.1463-1326.2008.00931.x 75

[499] Lecka-Czernik, B. *Bone loss in diabetes: use of antidiabetic thiazolidinediones and secondary osteoporosis*. Curr Osteoporos Rep 2010. 8:178–84. DOI: 10.1007/s11914-010-0027-y 75

[500] Goodman, WG. *Renal osteodystrophy*, in Principles of bone biology, J.P. Bilezikian, L.G. Raisz, and T.J. Martin, Editors. 2008, Academic Press/Elsevier: San Diego, CA. p. 1479–1510. 75

[501] Ott, SM. *Review article: Bone density in patients with chronic kidney disease stages 4–5.* Nephrology (Carlton) 2009. 14:395-403. DOI: 10.1111/j.1440-1797.2009.01159.x 76

[502] Levey, AS, Astor, BC, et al. *Chronic kidney disease, diabetes, and hypertension: what's in a name?* Kidney Int 2010. 78:19–22. DOI: 10.1038/ki.2010.115 76

[503] Herrero-Beaumont, G, Roman-Blas, JA, et al. *Primary osteoarthritis no longer primary: three subsets with distinct etiological, clinical, and therapeutic characteristics.* Semin Arthritis Rheum 2009. 39:71–80. DOI: 10.1016/j.semarthrit.2009.03.006 76

[504] Sowers, MR, McConnell, D, et al. *Estradiol and its metabolites and their association with knee osteoarthritis.* Arthritis Rheum 2006. 54:2481–7. DOI: 10.1002/art.22005 76

[505] Shane Anderson, A and Loeser, RF. *Why is osteoarthritis an age-related disease?* Best Pract Res Clin Rheumatol 2010. 24:15–26. DOI: 10.1016/j.berh.2009.08.006 76

[506] Meachim, G. *Light microscopy of Indian ink preparations of fibrillated cartilage.* Ann Rheum Dis 1972. 31:457–64. DOI: 10.1136/ard.31.6.457 77

[507] Rubin, E and Reisner, HM, Essentials of Rubin's pathology. 5th ed. 2009, Philadelphia: Wolters Kluwer/Lippincott Williams & Wilkins Health. p. 77

[508] Radin, EL and Rose, RM. *Role of subchondral bone in the initiation and progression of cartilage damage.* Clin Orthop Relat Res 1986. 34–40. 77

[509] Valdes, AM and Spector, TD. *The clinical relevance of genetic susceptibility to osteoarthritis.* Best Pract Res Clin Rheumatol 2010. 24:3–14. DOI: 10.1016/j.berh.2009.08.005 77

[510] Valdes, AM, Loughlin, J, et al. *Genome-wide association scan identifies a prostaglandin-endoperoxide synthase 2 variant involved in risk of knee osteoarthritis.* Am J Hum Genet 2008. 82:1231–40. DOI: 10.1016/j.ajhg.2008.04.006 77

[511] Martin, JA, Brown, TD, et al. *Chondrocyte senescence, joint loading and osteoarthritis.* Clin Orthop Relat Res 2004. S96–103. DOI: 10.1097/01.blo.0000143818.74887.b1 78

[512] Vignon, E and Arlot, M. *Macroscopically normal cartilage from the human osteoarthritic femoral head. II. Measurement of cartilage thickness and cell density.* J Rheumatol 1981. 8:447–50. 78

[513] Kouri, JB, Aguilera, JM, et al. *Apoptotic chondrocytes from osteoarthrotic human articular cartilage and abnormal calcification of subchondral bone.* J Rheumatol 2000. 27:1005–19. 78

[514] Chen, MH, Wang, JL, et al. *Relationship of chondrocyte apoptosis to matrix degradation and swelling potential of osteoarthritic cartilage.* J Formos Med Assoc 2005. 104:264–72. 78

[515] Goldring, MB and Goldring, SR. *Articular cartilage and subchondral bone in the pathogenesis of osteoarthritis.* Ann N Y Acad Sci 2010. 1192:230–7.
DOI: 10.1111/j.1749-6632.2009.05240.x 78

[516] Mankin, HJ, Johnson, ME, et al. *Biochemical and metabolic abnormalities in articular cartilage from osteoarthritic human hips. III. Distribution and metabolism of amino sugar-containing macromolecules.* J Bone Joint Surg Am 1981. 63:131–9. 78

[517] Palmoski, M and Brandt, K. *Hyaluronate-binding by proteoglycans. Comparison of mildly and severely osteoarthritic regions of human femoral cartilage.* Clin Chim Acta 1976. 70:87–95.
DOI: 10.1016/0009-8981(76)90008-5 78

[518] Martel-Pelletier, J, Boileau, C, et al. *Cartilage in normal and osteoarthritis conditions.* Best Pract Res Clin Rheumatol 2008. 22:351–84. DOI: 10.1016/j.berh.2008.02.001 78

[519] Outerbridge, RE. *The etiology of chondromalacia patellae.* J Bone Joint Surg Br 1961. 43-B:752–7. 78, 109

[520] Mankin, HJ, Dorfman, H, et al. *Biochemical and metabolic abnormalities in articular cartilage from osteo-arthritic human hips. II. Correlation of morphology with biochemical and metabolic data.* J Bone Joint Surg Am 1971. 53:523–37. 78

[521] Garvican, ER, Vaughan-Thomas, A, et al. *Biomarkers of cartilage turnover. Part 2: Non-collagenous markers.* Vet J 2010. 185:43–9. DOI: 10.1016/j.tvjl.2010.04.011 78

[522] Cawston, TE and Wilson, AJ. *Understanding the role of tissue degrading enzymes and their inhibitors in development and disease.* Best Pract Res Clin Rheumatol 2006. 20:983–1002.
DOI: 10.1016/j.berh.2006.06.007 78

[523] Pelletier, JP, Mineau, F, et al. *Imbalance between the mechanisms of activation and inhibition of metalloproteases in the early lesions of experimental osteoarthritis.* Arthritis Rheum 1990. 33:1466–76. DOI: 10.1002/art.1780331003 78

[524] Vittorio, N, Crissman, JD, et al. *Histologic assessment of cathepsin D in osteoarthritic cartilage.* Clin Exp Rheumatol 1986. 4:221–30. 78

[525] Shinmei, M, Masuda, K, et al. *Production of cytokines by chondrocytes and its role in proteoglycan degradation.* J Rheumatol Suppl 1991. 27:89–91. 78

[526] Foss, MV and Byers, PD. *Bone density, osteoarthrosis of the hip, and fracture of the upper end of the femur.* Ann Rheum Dis 1972. 31:259–64. DOI: 10.1136/ard.31.4.259 79

[527]   Knight, SM, Ring, EF, et al. *Bone mineral density and osteoarthritis*. Ann Rheum Dis 1992. 51:1025–6. DOI: 10.1136/ard.51.9.1025 79

[528]   Glowacki, J, Hurwitz, S, et al. *Osteoporosis and vitamin-D deficiency among postmenopausal women with osteoarthritis undergoing total hip arthroplasty*. J Bone Joint Surg Am 2003. 85-A:2371–7. 79

[529]   Sah, AP, Thornhill, TS, et al. *Correlation of plain radiographic indices of the hip with quantitative bone mineral density*. Osteoporos Int 2007. 18:1119–26. DOI: 10.1007/s00198-007-0348-6 79

[530]   Glowacki, J, Tuteja, M, et al. *Discordance in femoral neck bone density in subjects with unilateral hip osteoarthritis*. J Clin Densitom 2010. 13:24–8. DOI: 10.1016/j.jocd.2009.09.007 79

[531]   Setty, N, Rinaldi, G, et al. *Underestimated fracture risk in patients with unilateral hip osteoarthritis as calculated by FRAX*. J Bone Mineral Res 2010. 25:S115. 79

[532]   Smith, HR (2010) *Rheumatoid Arthritis*. eMedicine. Available at: `http://emedicine.medscape.com/article/331715-overview`. 79, 80, 81

[533]   *Rheumatoid Arthritis*. Published by: Mayo Clinic. Last Updated: 2010. Accessed on: November 2010. Available at: `http://www.mayoclinic.com/health/rheumatoid-arthritis/DS00020`. 79

[534]   Angeles-Han, S and Prahalad, S. *The genetics of juvenile idiopathic arthritis: what is new in 2010?* Curr Rheumatol Rep 2010. 12:87–93. DOI: 10.1007/s11926-010-0087-0 79

[535]   Scott, DL, Wolfe, F, et al. *Rheumatoid arthritis*. Lancet 2010. 376:1094–108. DOI: 10.1016/S0140-6736(10)60826-4 80, 81

[536]   Rantapaa-Dahlqvist, S, de Jong, BA, et al. *Antibodies against cyclic citrullinated peptide and IgA rheumatoid factor predict the development of rheumatoid arthritis*. Arthritis Rheum 2003. 48:2741–9. DOI: 10.1002/art.11223 80

[537]   Holick, MF. *The vitamin D deficiency pandemic and consequences for nonskeletal health: mechanisms of action*. Mol Aspects Med 2008. 29:361–8. DOI: 10.1016/j.mam.2008.08.008 80

[538]   Mathieu, C and Adorini, L. *The coming of age of 1,25-dihydroxyvitamin D(3) analogs as immunomodulatory agents*. Trends Mol Med 2002. 8:174–9. DOI: 10.1016/S1471-4914(02)02294-3 80

[539]   Andjelkovic, Z, Vojinovic, J, et al. *Disease modifying and immunomodulatory effects of high dose 1 alpha (OH) D3 in rheumatoid arthritis patients*. Clin Exp Rheumatol 1999. 17:453–6. 80

[540] Symmons, D, Turner, G, et al. *The prevalence of rheumatoid arthritis in the United Kingdom: new estimates for a new century*. Rheumatology (Oxford) 2002. 41:793–800. DOI: 10.1093/rheumatology/41.7.793 80

[541] Munz, C, Lunemann, JD, et al. *Antiviral immune responses: triggers of or triggered by autoimmunity?* Nat Rev Immunol 2009. 9:246–58. DOI: 10.1038/nri2527 80

[542] Feldmann, M, Brennan, FM, et al. *Rheumatoid arthritis*. Cell 1996. 85:307–10. DOI: 10.1016/S0092-8674(00)81109-5 80

[543] Choy, EH, Isenberg, DA, et al. *Therapeutic benefit of blocking interleukin-6 activity with an anti-interleukin-6 receptor monoclonal antibody in rheumatoid arthritis: a randomized, double-blind, placebo-controlled, dose-escalation trial*. Arthritis Rheum 2002. 46:3143–50. DOI: 10.1002/art.10623 80

[544] Meune, C, Touze, E, et al. *Trends in cardiovascular mortality in patients with rheumatoid arthritis over 50 years: a systematic review and meta-analysis of cohort studies*. Rheumatology (Oxford) 2009. 48:1309–13. DOI: 10.1093/rheumatology/kep252 81

[545] Kremer, JM and Lee, JK. *The safety and efficacy of the use of methotrexate in long-term therapy for rheumatoid arthritis*. Arthritis Rheum 1986. 29:822–31. DOI: 10.1002/art.1780290702 81

[546] O'Dell, JR, Haire, CE, et al. *Treatment of rheumatoid arthritis with methotrexate alone, sulfasalazine and hydroxychloroquine, or a combination of all three medications*. N Engl J Med 1996. 334:1287–91. DOI: 10.1056/NEJM199605163342002 81

[547] Maini, RN, Breedveld, FC, et al. *Therapeutic efficacy of multiple intravenous infusions of anti-tumor necrosis factor alpha monoclonal antibody combined with low-dose weekly methotrexate in rheumatoid arthritis*. Arthritis Rheum 1998. 41:1552–63. DOI: 10.1002/1529-0131(199809)41:9%3C1552::AID-ART5%3E3.0.CO;2-W 81

[548] McInnes, IB and O'Dell, JR. *State-of-the-art: rheumatoid arthritis*. Ann Rheum Dis 2010. 69:1898–906. DOI: 10.1136/ard.2010.134684 81

[549] Bui-Mansfield, LT (2009) *Osteochondritis Dissecans*. eMedicine. Available at: http://emedicine.medscape.com/article/392396-overview. 81, 82

[550] *Osteochondritis Dissecans*. Published by: Mayo Clinic. Last Updated: 2010. Accessed on: November 2010. Available at: http://www.mayoclinic.com/health/osteochondritis-dissecans/DS00741. 81

[551] Gomoll, AH, Farr, J, et al. *Surgical management of articular cartilage defects of the knee*. J Bone Joint Surg Am 2010. 92:2470–90. 82

[552] Tanaka, E, Detamore, MS, et al. *Degenerative disorders of the temporomandibular joint: etiology, diagnosis, and treatment*. J Dent Res 2008. 87:296–307.
DOI: 10.1177/154405910808700406 82, 83

[553] August, M and Glowacki, J. *Temporomandibular joint syndrome*, in Primary care of women, K.J. Carlson and S.A. Eisenstat, Editors. 2002, Mosby: St. Louis. p. 66–70. 82

[554] *TMJ Disorders*. Published by: Mayo Clinic. Last Updated: 2010. Accessed on: November 2010. Available at: http://www.mayoclinic.com/health/tmj-disorders/DS00355.
82

[555] Poveda Roda, R, Bagan, JV, et al. *Review of temporomandibular joint pathology. Part I: classification, epidemiology and risk factors*. Med Oral Patol Oral Cir Bucal 2007. 12:E292–8.
82

[556] Fricton, J, Look, JO, et al. *Systematic review and meta-analysis of randomized controlled trials evaluating intraoral orthopedic appliances for temporomandibular disorders*. J Orofac Pain 2010. 24:237–54. 82

[557] List, T and Axelsson, S. *Management of TMD: evidence from systematic reviews and meta-analyses*. J Oral Rehabil 2010. 37:430–51. DOI: 10.1111/j.1365-2842.2010.02089.x 82, 83

[558] Hersh, EV, Balasubramaniam, R, et al. *Pharmacologic management of temporomandibular disorders*. Oral Maxillofac Surg Clin North Am 2008. 20:197–210, vi.
DOI: 10.1016/j.coms.2007.12.005 83

[559] Ware, JC. *Tricyclic antidepressants in the treatment of insomnia*. J Clin Psychiatry 1983. 44:25–8. 83

[560] Manfredini, D, Piccotti, F, et al. *Hyaluronic acid in the treatment of TMJ disorders: a systematic review of the literature*. Cranio 2010. 28:166–76. 83

[561] La Touche, R, Goddard, G, et al. *Acupuncture in the treatment of pain in temporomandibular disorders: a systematic review and meta-analysis of randomized controlled trials*. Clin J Pain 2010. 26:541–50. DOI: 10.1097/AJP.0b013e3181e2697e 83

[562] Allen, KD and Athanasiou, KA. *Tissue Engineering of the TMJ disc: a review*. Tissue Eng 2006. 12:1183–96. DOI: 10.1089/ten.2006.12.1183 83

[563] Athanasiou, KA, Almarza, A, et al., Tissue Engineering of Temporomandibular Joint Cartilage, in Synthesis Lectures on Tissue Engineering, K.A. Athanasiou, Editor. 2009, Morgan & Claypool: San Rafael, CA. 83, 93

[564] Jansen, JA. *Animals models for studying soft tissue biocompatibility of biomaterials*, in Animal models in orthopaedic research, Y.H. An and R.J. Friedman, Editors. 1999, CRC Press: Boca Raton. p. 393–405. 85

[565] Glowacki, J. *Cellular Responses to Bone-Derived Materials*, in Bone and Cartilage Allografts: Biology and Clinical Applications, G. Friedlaender and V.M. Goldberg, Editors. 1991, American Academy of Orthopedic Surgeons. p. 55–73. 85

[566] Glowacki, J. *Inflammation and bone formation*, in Bone formation and repair, C.T. Brighton, G.E. Friedlaender, and J.M. Lane, Editors. 1994, American Academy of Orthopaedic Surgeons: Rosemont, IL. p. 395–403. 85, 86

[567] An, YH, Woolf, SK, et al. *Pre-clinical in vivo evaluation of orthopaedic bioabsorbable devices.* Biomaterials 2000. 21:2635–52. DOI: 10.1016/S0142-9612(00)00132-0 85

[568] Goodman, S, Aspenberg, P, et al. *Cement particles inhibit bone growth into titanium chambers implanted in the rabbit.* Acta Orthop Scand 1993. 64:627–33. DOI: 10.3109/17453679308994585 86

[569] Wang, JS, Goodman, S, et al. *Bone formation in the presence of phagocytosable hydroxyapatite particles.* Clin Orthop Relat Res 1994. 272–9. 86

[570] Bonnarens, F and Einhorn, TA. *Production of a standard closed fracture in laboratory animal bone.* J Orthop Res 1984. 2:97–101. DOI: 10.1002/jor.1100020115 86

[571] Hillier, ML and Bell, LS. *Differentiating human bone from animal bone: a review of histological methods.* J Forensic Sci 2007. 52:249–63. DOI: 10.1111/j.1556-4029.2006.00368.x 86, 88

[572] Goshima, J, Goldberg, VM, et al. *The origin of bone formed in composite grafts of porous calcium phosphate ceramic loaded with marrow cells.* Clin Orthop Relat Res 1991. 274–83. 89, 108

[573] Xia, Z, Ye, H, et al. *Efficient characterisation of human cell–bioceramic interactions in vitro and in vivo by using enhanced GFP-labelled mesenchymal stem cells.* Biomaterials 2005. 26:5790–800. DOI: 10.1016/j.biomaterials.2005.02.018 89

[574] Olivo, C, Alblas, J, et al. *In vivo bioluminescence imaging study to monitor ectopic bone formation by luciferase gene marked mesenchymal stem cells.* J Orthop Res 2008. 26:901–9. DOI: 10.1002/jor.20582 89

[575] Hannouche, D, Raould, A, et al. *Embedding of bone samples in methylmethacrylate: a suitable method for tracking LacZ mesenchymal stem cells in skeletal tissues.* J Histochem Cytochem 2007. 55:255–62. DOI: 10.1369/jhc.6A7063.2006 88, 89

[576] Iris, B, Zilberman, Y, et al. *Molecular imaging of the skeleton: quantitative real-time bioluminescence monitoring gene expression in bone repair and development.* J Bone Miner Res 2003. 18:570–8. DOI: 10.1359/jbmr.2003.18.3.570 88, 89

[577] de Boer, J, van Blitterswijk, C, et al. *Bioluminescent imaging: emerging technology for non-invasive imaging of bone tissue engineering.* Biomaterials 2006. 27:1851–8. DOI: 10.1016/j.biomaterials.2005.09.034 89

[578] Zhang, X, Xie, C, et al. *Periosteal progenitor cell fate in segmental cortical bone graft transplantations: implications for functional tissue engineering.* J Bone Miner Res 2005. 20:2124–37. DOI: 10.1359/JBMR.050806 89

[579] Duan, X, Yang, L, et al. *Characterization of EGFP-labeled mesenchymal stem cells and redistribution of allogeneic cells after subcutaneous implantation.* Arch Orthop Trauma Surg 2008. 128:751–9. DOI: 10.1007/s00402-008-0585-y 89

[580] Geuze, RE, Prins, HJ, et al. *Luciferase labeling for multipotent stromal cell tracking in spinal fusion versus ectopic bone tissue engineering in mice and rats.* Tissue Eng Part A 2010. 16:3343–51. DOI: 10.1089/ten.tea.2009.0774 89

[581] Huggins, CB. *The formation of bone under the influence of epithelium of the urinary tract.* Clin Orthop Relat Res 1968. 59:7–19. DOI: 10.1097/00003086-196807000-00002 90

[582] Sampath, TK and Reddi, AH. *Dissociative extraction and reconstitution of extracellular matrix components involved in local bone differentiation.* Proc Natl Acad Sci U S A 1981. 78:7599–603. 90, 103

[583] Mulliken, JB and Glowacki, J. *Induced osteogenesis for repair and construction in the craniofacial region.* Plast Reconstr Surg 1980. 65:553–60. DOI: 10.1097/00006534-198005000-00001 90, 91, 99

[584] Glowacki, J, Altobelli, D, et al. *Fate of mineralized and demineralized osseous implants in cranial defects.* Calcif Tissue Int 1981. 33:71–6. DOI: 10.1007/BF02409414 90

[585] Kaban, LB and Glowacki, J. *Induced osteogenesis in the repair of experimental mandibular defects in rats.* J Dent Res 1981. 60:1356–64. DOI: 10.1177/00220345810600071201 90

[586] Schmitz, JP and Hollinger, JO. *The critical size defect as an experimental model for craniomandibulofacial nonunions.* Clin Orthop Relat Res 1986. 299–308. 90

[587] Goodship, AE, Lanyon, LE, et al. *Functional adaptation of bone to increased stress. An experimental study.* J Bone Joint Surg Am 1979. 61:539–46. 90

[588] Holy, CE, Fialkov, MS, et al. *In Vivo Models for bone tissue-engineering constructs*, in Bone Engineering, J.E. Davies, Editor. 2000, EM squared incorporated: Toronto. p. 496–504. 90

[589] Kaban, LB and Glowacki, J. *Augmentation of rat mandibular ridge with demineralized bone implants.* J Dent Res 1984. 63:998–1002. DOI: 10.1177/00220345840630071801 90, 92

[590] Muschler, GF, Raut, VP, et al. *The design and use of animal models for translational research in bone tissue engineering and regenerative medicine.* Tissue Eng Part B Rev 2010. 16:123–45. DOI: 10.1089/ten.teb.2009.0658 90

[591] van der Worp, HB, Howells, DW, et al. *Can animal models of disease reliably inform human studies?* PLoS Med 2010. 7:e1000245. DOI: 10.1371/journal.pmed.1000245 91

[592] Osyczka, AM, Diefenderfer, DL, et al. *Different effects of BMP-2 on marrow stromal cells from human and rat bone.* Cells Tissues Organs 2004. 176:109–19. DOI: 10.1159/000075032 91

[593] Osyczka, AM and Leboy, PS. *Bone morphogenetic protein regulation of early osteoblast genes in human marrow stromal cells is mediated by extracellular signal-regulated kinase and phosphatidylinositol 3-kinase signaling.* Endocrinology 2005. 146:3428–37. DOI: 10.1210/en.2005-0303 91

[594] *Evaluation criteria for musculoskeletal and craniofacial tissue engineering constructs: a conference report.* Tissue Eng Part A 2008. 14:2089–104. DOI: 10.1089/ten.tea.2007.0383 91, 125

[595] van Weerden, WM and Romijn, JC. *Use of nude mouse xenograft models in prostate cancer research.* Prostate 2000. 43:263–71. DOI: 10.1002/1097-0045(20000601)43:4%3C263::AID-PROS5%3E3.0.CO;2-I 91

[596] Aspenberg, P and Andolf, E. *Bone induction by fetal and adult human bone matrix in athymic rats.* Acta Orthop Scand 1989. 60:195–9. DOI: 10.3109/17453678909149253 91

[597] Mashiko, S, Ishihara, A, et al. *A pair-feeding study reveals that a Y5 antagonist causes weight loss in diet-induced obese mice by modulating food intake and energy expenditure.* Mol Pharmacol 2007. 71:602–8. DOI: 10.1124/mol.106.029991 92

[598] Russell, JC, Proctor, SD, et al. *Pair feeding-mediated changes in metabolism: stress response and pathophysiology in insulin-resistant, atherosclerosis-prone JCR:LA-cp rats.* Am J Physiol Endocrinol Metab 2008. 294:E1078–87. DOI: 10.1152/ajpendo.90257.2008 92

[599] Shimomura, Y, Shimizu, H, et al. *Importance of feeding time in pair-fed, ovariectomized rats.* Physiol Behav 1989. 45:1197–200. DOI: 10.1016/0031-9384(89)90109-1 92

[600] *Aged Rodent Colonies Handbook.* Published by: National Institutes on Aging. Last Updated: 2010. Accessed on: December 17, 2010. Available at: http://www.nia.nih.gov/ResearchInformation/ScientificResources/AgedRodentColoniesHandbook. 92

[601] Tesch, GH and Allen, TJ. *Rodent models of streptozotocin-induced diabetic nephropathy.* Nephrology (Carlton) 2007. 12:261–6. DOI: 10.1111/j.1440-1797.2007.00796.x 92

[602] Lenzen, S. *The mechanisms of alloxan- and streptozotocin-induced diabetes.* Diabetologia 2008. 51:216–26. DOI: 10.1007/s00125-007-0886-7 92

[603] Andrikopoulos, S, Massa, CM, et al. *Differential effect of inbred mouse strain (C57BL/6, DBA/2, 129T2) on insulin secretory function in response to a high fat diet.* J Endocrinol 2005. 187:45–53. DOI: 10.1677/joe.1.06333 92

[604] Connor, JL. *Genetic mechanisms controlling the domestication of a wild house mouse population (Mus musculus L.).* J Comp Physiol Psychol 1975. 89:118–30. DOI: 10.1037/h0076656 92

[605] Stockwell, RA and Meachim, G. *The chondrocytes,* in Adult articular cartilage, M.A.R. Freeman, Editor. 1979, Pitman Medical: Tunbridge Wells, Eng. p. 69–144. 93

[606] Martino, F, Ettorre, GC, et al. *Articular cartilage echography as a criterion of the evolution of osteoarthritis of the knee.* Int J Clin Pharmacol Res 1993. 13 Suppl:35–42. 93

[607] An, YH and Friedman, RJ. *Animal models of articular cartilage defect,* in Animal models in orthopaedic research, Y.H. An and R.J. Friedman, Editors. 1999, CRC Press: Boca Raton. p. 309–325. 93

[608] Brandt, KD. *Animal models: insights into osteoarthritis (OA) provided by the cruciate-deficient dog.* Br J Rheumatol 1991. 30 Suppl 1:5–9. 93

[609] Pritzker, KP. *Animal models for osteoarthritis: processes, problems and prospects.* Ann Rheum Dis 1994. 53:406–20. DOI: 10.1136/ard.53.6.406 93

[610] Blaney Davidson, EN, Vitters, EL, et al. *Expression of transforming growth factor-beta (TGF-beta) and the TGFbeta signalling molecule SMAD-2P in spontaneous and instability-induced osteoarthritis: role in cartilage degradation, chondrogenesis and osteophyte formation.* Ann Rheum Dis 2006. 65:1414–21. DOI: 10.1136/ard.2005.045971 93

[611] Glasson, SS. *In vivo osteoarthritis target validation utilizing genetically-modified mice.* Curr Drug Targets 2007. 8:367–76. DOI: 10.2174/138945007779940061 93

[612] Li, Y, Xu, L, et al. *Lessons from genetic forms of osteoarthritis for the pathogenesis of the disease.* Osteoarthritis Cartilage 2007. 15:1101–5. DOI: 10.1016/j.joca.2007.04.013 93

[613] Di Paola, R and Cuzzocrea, S. *Predictivity and sensitivity of animal models of arthritis.* Autoimmun Rev 2008. 8:73–5. DOI: 10.1016/j.autrev.2008.07.029 93

[614] Herring, SW and Liu, ZJ. *Loading of the temporomandibular joint: anatomical and in vivo evidence from the bones.* Cells Tissues Organs 2001. 169:193–200. DOI: 10.1159/000047882 93

[615] Meikle, MC. *Remodeling the dentofacial skeleton: the biological basis of orthodontics and dentofacial orthopedics.* J Dent Res 2007. 86:12–24. DOI: 10.1177/154405910708600103 93

[616] Troulis, MJ, Glowacki, J, et al. *Effects of latency and rate on bone formation in a porcine mandibular distraction model.* J Oral Maxillofac Surg 2000. 58:507–13; discussion 514. DOI: 10.1016/S0278-2391(00)90012-0 93

[617] Thurmuller, P, Troulis, MJ, et al. *Microscopic changes in the condyle and disc in response to distraction osteogenesis of the minipig mandible.* J Oral Maxillofac Surg 2006. 64:249–58. DOI: 10.1016/j.joms.2005.10.010 93

[618] Balazs, EA. *Analgesic effect of elastoviscous hyaluronan solutions and the treatment of arthritic pain.* Cells Tissues Organs 2003. 174:49–62. DOI: 10.1159/000070574 94

[619] Smith, MH, Flanagan, CL, et al. *Computed tomography-based tissue-engineered scaffolds in craniomaxillofacial surgery.* Int J Med Robot 2007. 3:207–16. DOI: 10.1002/rcs.143 94

[620] Meacham, G. *The effect of scarification on articular cartilage in the rabbit.* J Bone Joint Surg 1963. 45B:150–161. 94

[621] Green, WT, Jr. *Articular cartilage repair. Behavior of rabbit chondrocytes during tissue culture and subsequent allografting.* Clin Orthop Relat Res 1977. 237–50. 94

[622] Grande, DA, Pitman, MI, et al. *The repair of experimentally produced defects in rabbit articular cartilage by autologous chondrocyte transplantation.* J Orthop Res 1989. 7:208–18. DOI: 10.1002/jor.1100070208 94

[623] Brittberg, M, Lindahl, A, et al. *Treatment of deep cartilage defects in the knee with autologous chondrocyte transplantation.* N Engl J Med 1994. 331:889–95. DOI: 10.1056/NEJM199410063311401 94, 109

[624] Upton, J, Sohn, SA, et al. *Neocartilage derived from transplanted perichondrium: what is it?* Plast Reconstr Surg 1981. 68:166–74. DOI: 10.1097/00006534-198108000-00007 94

[625] Ohlsen, L, Skoog, T, et al. *The pathogenesis of cauliflower ear. An experimental study in rabbits.* Scand J Plast Reconstr Surg 1975. 9:34–9. DOI: 10.3109/02844317509022854 94

[626] O'Driscoll, SW and Salter, RB. *The induction of neochondrogenesis in free intra-articular periosteal autografts under the influence of continuous passive motion. An experimental investigation in the rabbit.* J Bone Joint Surg Am 1984. 66:1248–57. 94

[627] Angermann, P, Riegels-Nielsen, P, et al. *Osteochondritis dissecans of the femoral condyle treated with periosteal transplantation. Poor outcome in 14 patients followed for 6–9 years.* Acta Orthop Scand 1998. 69:595–7. DOI: 10.3109/17453679808999262 94

[628] Madsen, BL, Noer, HH, et al. *Long-term results of periosteal transplantation in osteochondritis dissecans of the knee.* Orthopedics 2000. 23:223–6. 94

[629] Steinert, AF, Noth, U, et al. *Concepts in gene therapy for cartilage repair.* Injury 2008. 39 Suppl 1:S97–113. DOI: 10.1016/j.injury.2008.01.034 94

[630] Tortelli, F and Cancedda, R. *Three-dimensional cultures of osteogenic and chondrogenic cells: a tissue engineering approach to mimic bone and cartilage in vitro.* Eur Cell Mater 2009. 17:1–14. 94

[631] Kandel, R, Roberts, S, et al. *Tissue engineering and the intervertebral disc: the challenges.* Eur Spine J 2008. 17 Suppl 4:480–91. DOI: 10.1007/s00586-008-0746-2 94

[632] Bueno, EM and Glowacki, J. *Cell-free and cell-based approaches for bone regeneration.* Nat Rev Rheumatol 2009. 5:685–97. DOI: 10.1038/nrrheum.2009.228 94, 96, 99, 100, 125

[633] Rosenbaum, AJ, Grande, DA, et al. *The use of mesenchymal stem cells in tissue engineering: A global assessment.* Organogenesis 2008. 4:23–7. 94

[634] Glowacki, J. *Histogenesis,* in Encyclopedia of biomaterials and biomedical engineering, G.E. Wnek and G.L. Bowlin, Editors. 2004, Marcel Dekker: New York. p. 762–769. 95

[635] Kolar, P, Schmidt-Bleek, K, et al. *The early fracture hematoma and its potential role in fracture healing.* Tissue Eng Part B Rev 2010. 16:427–34. DOI: 10.1089/ten.teb.2009.0687 95

[636] Temenoff, JS, Lu, L, et al. *Bone tissue-engineering using synthetic biodegradable polymer scaffolds,* in Bone Engineering, J.E. Davies, Editor. 2000, EM squared incorporated: Toronto. p. 454–461. 96, 97

[637] Hutmacher, DW. *Scaffolds in tissue engineering bone and cartilage.* Biomaterials 2000. 21:2529–43. DOI: 10.1016/S0142-9612(00)00121-6 96

[638] Steele, J, Dalton, B, et al. *Underlying mechanisms of cellular adhesion in vitro during colonization of synthetic surfaces by bone-derived cells,* in Bone Engineering, J.E. Davies, Editor. 2000, EM Square: Toronto. p. 225–231. 96

[639] Laurencin, CT and Lu, HH. *Polymer-ceramic composites for bone tissue-engineering,* in Bone Engineering, J.E. Davies, Editor. 2000, EM squared incorporated: Toronto. p. 462–472. 97, 100

[640] Meyer, U, Joos, U, et al. *Biological and biophysical principles in extracorporal bone tissue engineering. Part I.* Int J Oral Maxillofac Surg 2004. 33:325–32. DOI: 10.1016/S0901-5027(03)00199-1 97

[641] Gauthier, O, Bouler, JM, et al. *Macroporous biphasic calcium phosphate ceramics: influence of macropore diameter and macroporosity percentage on bone ingrowth.* Biomaterials 1998. 19:133–9. DOI: 10.1016/S0142-9612(97)00180-4 97, 100

[642] Karageorgiou, V and Kaplan, D. *Porosity of 3D biomaterial scaffolds and osteogenesis.* Biomaterials 2005. 26:5474–91. DOI: 10.1016/j.biomaterials.2005.02.002 97

[643] Ripamonti, U, Ma, S, et al. *Initiation of bone regeneration in adult baboons by osteogenin, a bone morphogenetic protein.* Matrix 1992. 12:369–80. 97

[644] Buser, D, Schenk, RK, et al. *Influence of surface characteristics on bone integration of titanium implants. A histomorphometric study in miniature pigs.* J Biomed Mater Res 1991. 25:889–902. DOI: 10.1002/jbm.820250708 97

[645] Sitharaman, B, Shi, X, et al. *In vivo biocompatibility of ultra-short single-walled carbon nanotube/biodegradable polymer nanocomposites for bone tissue engineering.* Bone 2008. 43:362–70. DOI: 10.1016/j.bone.2008.04.013 97

[646] Webster, TJ, Siegel, RW, et al. *Osteoblast adhesion on nanophase ceramics.* Biomaterials 1999. 20:1221–7. DOI: 10.1016/S0142-9612(99)00020-4 97

[647] Moroni, L, de Wijn, JR, et al. *Integrating novel technologies to fabricate smart scaffolds.* J Biomater Sci Polym Ed 2008. 19:543–72. DOI: 10.1163/156856208784089571 98

[648] Woodfield, TB, Malda, J, et al. *Design of porous scaffolds for cartilage tissue engineering using a three-dimensional fiber-deposition technique.* Biomaterials 2004. 25:4149–61. DOI: 10.1016/j.biomaterials.2003.10.056 98

[649] Chu, TM, Orton, DG, et al. *Mechanical and in vivo performance of hydroxyapatite implants with controlled architectures.* Biomaterials 2002. 23:1283–93. DOI: 10.1016/S0142-9612(01)00243-5 98

[650] Lee, KW, Wang, S, et al. *Poly(propylene fumarate) bone tissue engineering scaffold fabrication using stereolithography: effects of resin formulations and laser parameters.* Biomacromolecules 2007. 8:1077–84. DOI: 10.1021/bm060834v

[651] Seitz, H, Rieder, W, et al. *Three-dimensional printing of porous ceramic scaffolds for bone tissue engineering.* J Biomed Mater Res B Appl Biomater 2005. 74:782–8. DOI: 10.1002/jbm.b.30291 98

[652] Kwon, IK, Kidoaki, S, et al. *Electrospun nano- to microfiber fabrics made of biodegradable copolyesters: structural characteristics, mechanical properties and cell adhesion potential.* Biomaterials 2005. 26:3929–39. DOI: 10.1016/j.biomaterials.2004.10.007 98

[653] Papenburg, BJ, Vogelaar, L, et al. *One-step fabrication of porous micropatterned scaffolds to control cell behavior.* Biomaterials 2007. 28:1998–2009. DOI: 10.1016/j.biomaterials.2006.12.023 98

[654] Hollister, SJ, Levy, RA, et al. *An image-based approach for designing and manufacturing craniofacial scaffolds.* Int J Oral Maxillofac Surg 2000. 29:67–71. DOI: 10.1016/S0901-5027(00)80128-9 98

[655] Burkoth, AK and Anseth, KS. *A review of photocrosslinked polyanhydrides: in situ forming degradable networks.* Biomaterials 2000. 21:2395–404. DOI: 10.1016/S0142-9612(00)00107-1 98

[656] Behravesh, E, Zygourakis, K, et al. *Adhesion and migration of marrow-derived osteoblasts on injectable in situ crosslinkable poly(propylene fumarate-co-ethylene glycol)-based hydrogels with a covalently linked RGDS peptide.* J Biomed Mater Res A 2003. 65:260–70. DOI: 10.1002/jbm.a.10461 98

[657] Burdick, JA, Frankel, D, et al. *An initial investigation of photocurable three-dimensional lactic acid based scaffolds in a critical-sized cranial defect.* Biomaterials 2003. 24:1613–20. DOI: 10.1016/S0142-9612(02)00538-0 98, 99

[658] Seeherman, HJ, Azari, K, et al. *rhBMP-2 delivered in a calcium phosphate cement accelerates bridging of critical-sized defects in rabbit radii.* J Bone Joint Surg Am 2006. 88:1553–65. DOI: 10.2106/JBJS.E.01006 98

[659] Link, DP, van den Dolder, J, et al. *Mechanical evaluation of implanted calcium phosphate cement incorporated with PLGA microparticles.* Biomaterials 2006. 27:4941–7. DOI: 10.1016/j.biomaterials.2006.05.022 98

[660] *Senn on the Healing of Aseptic Bone Cavities by Implantation of Antiseptic Decalcified Bone.* Ann Surg 1889. 10:352–68. DOI: 10.1097/00000658-188907000-00043 98

[661] Chen, B, Lin, H, et al. *Homogeneous osteogenesis and bone regeneration by demineralized bone matrix loading with collagen-targeting bone morphogenetic protein-2.* Biomaterials 2007. 28:1027–35. DOI: 10.1016/j.biomaterials.2006.10.013 99

[662] Clokie, CM and Sandor, GK. *Reconstruction of 10 major mandibular defects using bioimplants containing BMP-7.* J Can Dent Assoc 2008. 74:67–72. 99

[663] Han, B, Woodell-May, J, et al. *The effect of thrombin activation of platelet-rich plasma on demineralized bone matrix osteoinductivity.* J Bone Joint Surg Am 2009. 91:1459–70. DOI: 10.2106/JBJS.H.00246 99

[664] Lewin-Epstein, J. *Polyvinyl sponge (ivalon) as a scaffold for bone.* Br J Oral Surg 1964. 2:115–9. DOI: 10.1016/S0007-117X(64)80025-1 99

[665] Black, J and Hastings, G, Handbook of Biomaterial Properties. 1998, London: Chapman & Hall. 590. 99

[666] Lu, L, Peter, SJ, et al. *In vitro and in vivo degradation of porous poly(DL-lactic-co-glycolic acid) foams.* Biomaterials 2000. 21:1837–45. DOI: 10.1016/S0142-9612(00)00047-8 99

[667] James, K, Levene, H, et al. *Small changes in chemical structure of a polymer can have a significant effect on the hard-tissue response in vivo*, in Bone Engineering, J.E. Davies, Editor. 2000, EM squared incorporated: Toronto. p. 195–204. 99

[668] Murphy, WL, Simmons, CA, et al. *Bone regeneration via a mineral substrate and induced angiogenesis.* J Dent Res 2004. 83:204–10. DOI: 10.1177/154405910408300304 99

[669] Cowan, CM, Aghaloo, T, et al. *MicroCT evaluation of three-dimensional mineralization in response to BMP-2 doses in vitro and in critical sized rat calvarial defects.* Tissue Eng 2007. 13:501–12. DOI: 10.1089/ten.2006.0141 99

[670] Lee, SC, Shea, M, et al. *Healing of large segmental defects in rat femurs is aided by RhBMP-2 in PLGA matrix.* J Biomed Mater Res 1994. 28:1149–56. DOI: 10.1002/jbm.820281005 99

[671] Pitt, CG. *Poly-e-caprolactone and its copolymers*, in Biodegradable polymers as drug delivery systems, M. Chasin and R.S. Langer, Editors. 1990, M. Dekker: New York. p. 71–120. 99

[672] Piskin, E, Isoglu, IA, et al. *In vivo performance of simvastatin-loaded electrospun spiral-wound polycaprolactone scaffolds in reconstruction of cranial bone defects in the rat model.* J Biomed Mater Res A 2009. 90:1137–51. DOI: 10.1002/jbm.a.32157 99, 104

[673] Fisher, JP, Vehof, JW, et al. *Soft and hard tissue response to photocrosslinked poly(propylene fumarate) scaffolds in a rabbit model.* J Biomed Mater Res 2002. 59:547–56. DOI: 10.1002/jbm.1268 99

[674] Patel, ZS, Young, S, et al. *Dual delivery of an angiogenic and an osteogenic growth factor for bone regeneration in a critical size defect model.* Bone 2008. 43:931–40. DOI: 10.1016/j.bone.2008.06.019 99

[675] Braunecker, J, Baba, M, et al. *The effects of molecular weight and porosity on the degradation and drug release from polyglycolide.* Int J Pharm 2004. 282:19–34. DOI: 10.1016/j.ijpharm.2003.08.020 99

[676] Thomson, RC, Yaszemski, MJ, et al. *Fabrication of biodegradable polymer scaffolds to engineer trabecular bone.* J Biomater Sci Polym Ed 1995. 7:23–38. DOI: 10.1163/156856295X00805 99

[677] Zhang, R and Ma, PX. *Porous poly(L-lactic acid)/apatite composites created by biomimetic process.* J Biomed Mater Res 1999. 45:285–93. DOI: 10.1002/(SICI)1097-4636(19990615)45:4%3C285::AID-JBM2%3E3.3.CO;2-U 99

[678] Marcolongo, M, Ducheyne, P, et al. *Bioactive glass fiber/polymeric composites bond to bone tissue.* J Biomed Mater Res 1998. 39:161–70.
DOI: 10.1002/(SICI)1097-4636(199801)39:1%3C161::AID-JBM18%3E3.0.CO;2-I 100

[679] Chung, C and Burdick, JA. *Engineering cartilage tissue.* Adv Drug Deliv Rev 2008. 60:243–62. DOI: 10.1016/j.addr.2007.08.027 100, 111, 116, 117

[680] Al-Munajjed, AA, Plunkett, NA, et al. *Development of a biomimetic collagen-hydroxyapatite scaffold for bone tissue engineering using a SBF immersion technique.* J Biomed Mater Res B Appl Biomater 2009. 90:584–91. DOI: 10.1002/jbm.b.31320 100

[681] Hong, Y, Gong, Y, et al. *Collagen-coated polylactide microcarriers/chitosan hydrogel composite: injectable scaffold for cartilage regeneration.* J Biomed Mater Res A 2008. 85:628–37. DOI: 10.1002/jbm.a.31603 100

[682] Cancedda, R, Giannoni, P, et al. *A tissue engineering approach to bone repair in large animal models and in clinical practice.* Biomaterials 2007. 28:4240–50.
DOI: 10.1016/j.biomaterials.2007.06.023 100, 102

[683] White, AP, Vaccaro, AR, et al. *Clinical applications of BMP-7/OP-1 in fractures, nonunions and spinal fusion.* Int Orthop 2007. 31:735–41. DOI: 10.1007/s00264-007-0422-x 100, 102

[684] Carter, JD, Swearingen, AB, et al. *Clinical and radiographic assessment of transforaminal lumbar interbody fusion using HEALOS collagen-hydroxyapatite sponge with autologous bone marrow aspirate.* Spine J 2009. 9:434–8. DOI: 10.1016/j.spinee.2008.11.004 100

[685] Epstein, NE. *Beta tricalcium phosphate: observation of use in 100 posterolateral lumbar instrumented fusions.* Spine J 2009. 9:630–8. DOI: 10.1016/j.spinee.2009.04.007 100

[686] Hulbert, SF, Young, FA, et al. *Potential of ceramic materials as permanently implantable skeletal prostheses.* J Biomed Mater Res 1970. 4:433–56. DOI: 10.1002/jbm.820040309 100

[687] Ooms, EM, Wolke, JG, et al. *Trabecular bone response to injectable calcium phosphate (Ca-P) cement.* J Biomed Mater Res 2002. 61:9–18. DOI: 10.1002/jbm.10029 100

[688] Link, DP, van den Dolder, J, et al. *Bone response and mechanical strength of rabbit femoral defects filled with injectable CaP cements containing TGF-beta 1 loaded gelatin microparticles.* Biomaterials 2008. 29:675–82. DOI: 10.1016/j.biomaterials.2007.10.029 100

[689] Ruhe, PQ, Hedberg-Dirk, EL, et al. *Porous poly(DL-lactic-co-glycolic acid)/calcium phosphate cement composite for reconstruction of bone defects.* Tissue Eng 2006. 12:789–800.
DOI: 10.1089/ten.2006.12.789 100

[690] Schepers, E, de Clercq, M, et al. *Bioactive glass particulate material as a filler for bone lesions.* J Oral Rehabil 1991. 18:439–52. DOI: 10.1111/j.1365-2842.1991.tb01689.x 101

[691] Kokubo, T, Kim, HM, et al. *What kinds of materials exhibit bone-bonding?*, in Bone Engineering, J.E. Davies, Editor. 2000, EM squared incorporated: Toronto. p. 190–194. 101

[692] Hench, LL and Paschall, HA. *Direct chemical bond of bioactive glass-ceramic materials to bone and muscle.* J Biomed Mater Res 1973. 7:25–42. DOI: 10.1002/jbm.820070304 101

[693] Oonishi, H, Hench, LL, et al. *Quantitative comparison of bone growth behavior in granules of Bioglass, A-W glass-ceramic, and hydroxyapatite.* J Biomed Mater Res 2000. 51:37–46. DOI: 10.1002/(SICI)1097-4636(200007)51:1%3C37::AID-JBM6%3E3.0.CO;2-T 101

[694] Xue, W, Liu, X, et al. *In vivo evaluation of plasma-sprayed wollastonite coating.* Biomaterials 2005. 26:3455–60. DOI: 10.1016/j.biomaterials.2004.09.003 101

[695] Hedberg, EL, Kroese-Deutman, HC, et al. *Effect of varied release kinetics of the osteogenic thrombin peptide TP508 from biodegradable, polymeric scaffolds on bone formation in vivo.* J Biomed Mater Res A 2005. 72:343–53. DOI: 10.1002/jbm.a.30265 101, 104

[696] Babensee, JE, McIntire, LV, et al. *Growth factor delivery for tissue engineering.* Pharm Res 2000. 17:497–504. DOI: 10.1023/A:1007502828372 101

[697] Meijer, GJ, de Bruijn, JD, et al. *Cell-based bone tissue engineering.* PLoS Med 2007. 4:e9. DOI: 10.1371/journal.pmed.0040009 101, 108

[698] Evans, CH, Ghivizzani, SC, et al. *Orthopedic gene therapy in 2008.* Mol Ther 2009. 17:231–44. DOI: 10.1038/mt.2008.265 102

[699] Rundle, CH, Miyakoshi, N, et al. *In vivo bone formation in fracture repair induced by direct retroviral-based gene therapy with bone morphogenetic protein-4.* Bone 2003. 32:591–601. DOI: 10.1016/S8756-3282(03)00096-6 102

[700] Shea, LD, Smiley, E, et al. *DNA delivery from polymer matrices for tissue engineering.* Nat Biotechnol 1999. 17:551–4. DOI: 10.1038/9853 102

[701] Bonadio, J, Smiley, E, et al. *Localized, direct plasmid gene delivery in vivo: prolonged therapy results in reproducible tissue regeneration.* Nat Med 1999. 5:753–9. DOI: 10.1038/10473 102

[702] Fang, J, Zhu, YY, et al. *Stimulation of new bone formation by direct transfer of osteogenic plasmid genes.* Proc Natl Acad Sci U S A 1996. 93:5753–8. 102

[703] Reddi, AH. *Bone and cartilage differentiation.* Curr Opin Genet Dev 1994. 4:737–44. DOI: 10.1016/0959-437X(94)90141-O 102

[704] Cheng, H, Jiang, W, et al. *Osteogenic activity of the fourteen types of human bone morphogenetic proteins (BMPs).* J Bone Joint Surg Am 2003. 85-A:1544–52. 102

[705] Govender, S, Csimma, C, et al. *Recombinant human bone morphogenetic protein-2 for treatment of open tibial fractures: a prospective, controlled, randomized study of four hundred and fifty patients.* J Bone Joint Surg Am 2002. 84-A:2123–34. 102

[706] Vaccaro, AR, Whang, PG, et al. *The safety and efficacy of OP-1 (rhBMP-7) as a replacement for iliac crest autograft for posterolateral lumbar arthrodesis: minimum 4-year follow-up of a pilot study.* Spine J 2008. 8:457–65. DOI: 10.1016/j.spinee.2008.06.074 102

[707] Gautschi, OP, Frey, SP, et al. *Bone morphogenetic proteins in clinical applications.* ANZ J Surg 2007. 77:626–31. DOI: 10.1111/j.1445-2197.2007.04175.x 102

[708] Miyazaki, M, Morishita, Y, et al. *A porcine collagen-derived matrix as a carrier for recombinant human bone morphogenetic protein-2 enhances spinal fusion in rats.* Spine J 2009. 9:22–30. DOI: 10.1016/j.spinee.2008.08.009 102

[709] Jane, JA, Jr., Dunford, BA, et al. *Ectopic osteogenesis using adenoviral bone morphogenetic protein (BMP)-4 and BMP-6 gene transfer.* Mol Ther 2002. 6:464–70. DOI: 10.1006/mthe.2002.0691 102

[710] Chen, Y, Cheung, KM, et al. *In vivo new bone formation by direct transfer of adenoviral-mediated bone morphogenetic protein-4 gene.* Biochem Biophys Res Commun 2002. 298:121–7. DOI: 10.1016/S0006-291X(02)02394-X 102

[711] Betz, OB, Betz, VM, et al. *Delayed administration of adenoviral BMP-2 vector improves the formation of bone in osseous defects.* Gene Ther 2007. 14:1039–44. DOI: 10.1038/sj.gt.3302956 103

[712] Egermann, M, Lill, CA, et al. *Effect of BMP-2 gene transfer on bone healing in sheep.* Gene Ther 2006. 13:1290–9. DOI: 10.1038/sj.gt.3302785 103

[713] Huang, YC, Simmons, C, et al. *Bone regeneration in a rat cranial defect with delivery of PEI-condensed plasmid DNA encoding for bone morphogenetic protein-4 (BMP-4).* Gene Ther 2005. 12:418–26. DOI: 10.1038/sj.gt.3302439 103

[714] Bright, C, Park, YS, et al. *In vivo evaluation of plasmid DNA encoding OP-1 protein for spine fusion.* Spine (Phila Pa 1976) 2006. 31:2163–72. DOI: 10.1097/01.brs.0000232721.59901.45 103

[715] Hsu, WK and Wang, JC. *The use of bone morphogenetic protein in spine fusion.* Spine J 2008. 8:419–25. DOI: 10.1016/j.spinee.2008.01.008 103

[716] Zhou, S, Glowacki, J, et al. *Comparison of TGF-beta/BMP pathways signaled by demineralized bone powder and BMP-2 in human dermal fibroblasts.* J Bone Miner Res 2004. 19:1732–41. DOI: 10.1359/JBMR.040702 103

[717] Sampath, TK, Coughlin, JE, et al. *Bovine osteogenic protein is composed of dimers of OP-1 and BMP-2A, two members of the transforming growth factor-beta superfamily.* J Biol Chem 1990. 265:13198–205. 103

[718] Israel, DI, Nove, J, et al. *Heterodimeric bone morphogenetic proteins show enhanced activity in vitro and in vivo.* Growth Factors 1996. 13:291–300. DOI: 10.3109/08977199609003229 103

[719] Yanagita, M. *BMP antagonists: their roles in development and involvement in pathophysiology.* Cytokine Growth Factor Rev 2005. 16:309–17. DOI: 10.1016/j.cytogfr.2005.02.007 103

[720] Luo, X, Chen, J, et al. *Osteogenic BMPs promote tumor growth of human osteosarcomas that harbor differentiation defects.* Lab Invest 2008. 88:1264–77. DOI: 10.1038/labinvest.2008.98 103

[721] Lakey, LA, Akella, R, et al. *Angiogenesis: implications for tissue repair,* in Bone Engineering, J.E. Davies, Editor. 2000, EM squared incorporated: Toronto. p. 137–142. 103

[722] Ito, H, Koefoed, M, et al. *Remodeling of cortical bone allografts mediated by adherent rAAV-RANKL and VEGF gene therapy.* Nat Med 2005. 11:291–7. DOI: 10.1038/nm1190 103

[723] Geiger, F, Bertram, H, et al. *Vascular endothelial growth factor gene-activated matrix (VEGF165-GAM) enhances osteogenesis and angiogenesis in large segmental bone defects.* J Bone Miner Res 2005. 20:2028–35. DOI: 10.1359/JBMR.050701 103

[724] Geiger, F, Lorenz, H, et al. *VEGF producing bone marrow stromal cells (BMSC) enhance vascularization and resorption of a natural coral bone substitute.* Bone 2007. 41:516–22. DOI: 10.1016/j.bone.2007.06.018 103

[725] Ruoslahti, E and Pierschbacher, MD. *New perspectives in cell adhesion: RGD and integrins.* Science 1987. 238:491–7. DOI: 10.1126/science.2821619 103

[726] Garcia, AJ and Reyes, CD. *Bio-adhesive surfaces to promote osteoblast differentiation and bone formation.* J Dent Res 2005. 84:407–13. DOI: 10.1177/154405910508400502 103

[727] Eid, K, Chen, E, et al. *Effect of RGD coating on osteocompatibility of PLGA-polymer disks in a rat tibial wound.* J Biomed Mater Res 2001. 57:224–31. DOI: 10.1002/1097-4636(200111)57:2%3C224::AID-JBM1162%3E3.0.CO;2-F 103

[728] Shin, H, Jo, S, et al. *Modulation of marrow stromal osteoblast adhesion on biomimetic oligo[poly(ethylene glycol) fumarate] hydrogels modified with Arg-Gly-Asp peptides and a poly(ethyleneglycol) spacer.* J Biomed Mater Res 2002. 61:169–79. DOI: 10.1002/jbm.10193 104

[729] Burdick, JA and Anseth, KS. *Photoencapsulation of osteoblasts in injectable RGD-modified PEG hydrogels for bone tissue engineering.* Biomaterials 2002. 23:4315–23. DOI: 10.1016/S0142-9612(02)00176-X 104

[730] Dee, KC, Andersen, TT, et al. *Design and function of novel osteoblast-adhesive peptides for chemical modification of biomaterials.* J Biomed Mater Res 1998. 40:371–7. DOI: 10.1002/(SICI)1097-4636(19980605)40:3%3C371::AID-JBM5%3E3.0.CO;2-C 104

[731] Harbers, GM and Healy, KE. *The effect of ligand type and density on osteoblast adhesion, proliferation, and matrix mineralization.* J Biomed Mater Res A 2005. 75:855–69. DOI: 10.1002/jbm.a.30482 104

[732] Rosen, PS and Reynolds, MA. *A retrospective case series comparing the use of demineralized freeze-dried bone allograft and freeze-dried bone allograft combined with enamel matrix derivative for the treatment of advanced osseous lesions.* J Periodontol 2002. 73:942–9. DOI: 10.1902/jop.2002.73.8.942 104

[733] Yukna, RA and Mellonig, JT. *Histologic evaluation of periodontal healing in humans following regenerative therapy with enamel matrix derivative. A 10-case series.* J Periodontol 2000. 71:752–9. DOI: 10.1902/jop.2000.71.5.752 104

[734] Kawana, F, Sawae, Y, et al. *Porcine enamel matrix derivative enhances trabecular bone regeneration during wound healing of injured rat femur.* Anat Rec 2001. 264:438–46. DOI: 10.1002/ar.10016 104

[735] Savarino, L, Cenni, E, et al. *Evaluation of bone healing enhancement by lyophilized bone grafts supplemented with platelet gel: a standardized methodology in patients with tibial osteotomy for genu varus.* J Biomed Mater Res B Appl Biomater 2006. 76:364–72. DOI: 10.1002/jbm.b.30375 104

[736] Dallari, D, Savarino, L, et al. *Enhanced tibial osteotomy healing with use of bone grafts supplemented with platelet gel or platelet gel and bone marrow stromal cells.* J Bone Joint Surg Am 2007. 89:2413–20. DOI: 10.2106/JBJS.F.01026 104

[737] Choi, BH, Im, CJ, et al. *Effect of platelet-rich plasma on bone regeneration in autogenous bone graft.* Int J Oral Maxillofac Surg 2004. 33:56–9. DOI: 10.1054/ijom.2003.0466 104

[738] Yamada, Y, Ueda, M, et al. *Autogenous injectable bone for regeneration with mesenchymal stem cells and platelet-rich plasma: tissue-engineered bone regeneration.* Tissue Eng 2004. 10:955–64. DOI: 10.1089/1076327041348284 104

[739] Garrett, IR, Gutierrez, G, et al. *Statins and bone formation.* Curr Pharm Des 2001. 7:715–36. DOI: 10.2174/1381612013397762 104

[740] Wong, RW and Rabie, AB. *Statin collagen grafts used to repair defects in the parietal bone of rabbits*. Br J Oral Maxillofac Surg 2003. 41:244–8. DOI: 10.1016/S0266-4356(03)00081-0 104

[741] Friedenstein, AJ, Piatetzky, S, II, et al. *Osteogenesis in transplants of bone marrow cells*. J Embryol Exp Morphol 1966. 16:381–90. 104

[742] Pittenger, MF, Mackay, AM, et al. *Multilineage potential of adult human mesenchymal stem cells*. Science 1999. 284:143–7. DOI: 10.1126/science.284.5411.143 104

[743] Arthur, A, Zannettino, A, et al. *The therapeutic applications of multipotential mesenchymal/stromal stem cells in skeletal tissue repair*. J Cell Physiol 2009. 218:237–45. DOI: 10.1002/jcp.21592 104

[744] Lee, K, Chan, CK, et al. *Cell therapy for bone regeneration–bench to bedside*. J Biomed Mater Res B Appl Biomater 2009. 89:252–63.

[745] Sterodimas, A, De Faria, J, et al. *Tissue engineering in plastic surgery: an up-to-date review of the current literature*. Ann Plast Surg 2009. 62:97–103. DOI: 10.1097/SAP.0b013e3181788ec9 104

[746] Bianco, P, Robey, PG, et al. *"Mesenchymal" stem cells in human bone marrow (skeletal stem cells): a critical discussion of their nature, identity, and significance in incurable skeletal disease*. Hum Gene Ther 2010. 21:1057–66. DOI: 10.1089/hum.2010.136 104

[747] Caplan, AI. *What's in a name?* Tissue Eng Part A 2010. 16:2415–7. DOI: 10.1089/ten.tea.2010.0216 105, 125

[748] Gimble, J and Guilak, F. *Adipose-derived adult stem cells: isolation, characterization, and differentiation potential*. Cytotherapy 2003. 5:362–9. DOI: 10.1080/14653240310003026 105

[749] Crisan, M, Yap, S, et al. *A perivascular origin for mesenchymal stem cells in multiple human organs*. Cell Stem Cell 2008. 3:301–13. DOI: 10.1016/j.stem.2008.07.003 106

[750] Maherali, N, Ahfeldt, T, et al. *A high-efficiency system for the generation and study of human induced pluripotent stem cells*. Cell Stem Cell 2008. 3:340–5. DOI: 10.1016/j.stem.2008.08.003 106

[751] Le Blanc, K, Tammik, C, et al. *HLA expression and immunologic properties of differentiated and undifferentiated mesenchymal stem cells*. Exp Hematol 2003. 31:890–6. DOI: 10.1016/S0301-472X(03)00110-3 106

[752] Tasso, R, Augello, A, et al. *Recruitment of a host's osteoprogenitor cells using exogenous mesenchymal stem cells seeded on porous ceramic*. Tissue Eng Part A 2009. 15:2203–12. DOI: 10.1089/ten.tea.2008.0269 106, 125

[753] Parker, E, Shiga, A, et al. *human bone in vitro*, in Bone Engineering, J.E. Davies, Editor. 2000, EM Square: Toronto. p. 63–77. 106

[754] Mueller, SM and Glowacki, J. *Construction and regulation of 3-dimensional bone tissue in vitro*, in Bone Engineering, J.E. Davies, Editor. 2000, EM Square: Toronto. p. 473–483. 106

[755] Kuznetsov, SA, Krebsbach, PH, et al. *Single-colony derived strains of human marrow stromal fibroblasts form bone after transplantation in vivo*. J Bone Miner Res 1997. 12:1335–47. DOI: 10.1359/jbmr.1997.12.9.1335 107, 124

[756] Mueller, SM, Mizuno, S, et al. *Medium perfusion enhances osteogenesis by murine osteosarcoma cells in three-dimensional collagen sponges*. J Bone Miner Res 1999. 14:2118–26. DOI: 10.1359/jbmr.1999.14.12.2118 107

[757] Sikavitsas, VI, Bancroft, GN, et al. *Mineralized matrix deposition by marrow stromal osteoblasts in 3D perfusion culture increases with increasing fluid shear forces*. Proc Natl Acad Sci U S A 2003. 100:14683–8. DOI: 10.1073/pnas.2434367100 107

[758] Wiesmann, HP, Joos, U, et al. *Biological and biophysical principles in extracorporal bone tissue engineering. Part II.* Int J Oral Maxillofac Surg 2004. 33:523–30. DOI: 10.1016/j.ijom.2004.04.005 107

[759] Mikos, AG, Sarakinos, G, et al. *Prevascularization of porous biodegradable polymers.* Biotechnol Bioeng 1993. 42:716–23. DOI: 10.1002/bit.260420606 107

[760] Arkudas, A, Beier, JP, et al. *Axial prevascularization of porous matrices using an arteriovenous loop promotes survival and differentiation of transplanted autologous osteoblasts*. Tissue Eng 2007. 13:1549–60. DOI: 10.1089/ten.2006.0387 107

[761] Rouwkema, J, de Boer, J, et al. *Endothelial cells assemble into a 3-dimensional prevascular network in a bone tissue engineering construct*. Tissue Eng 2006. 12:2685–93. DOI: 10.1089/ten.2006.12.2685 107

[762] Taylor, D, Hazenberg, JG, et al. *Living with cracks: damage and repair in human bone*. Nat Mater 2007. 6:263–8. DOI: 10.1038/nmat1866 107

[763] Goshima, J, Goldberg, VM, et al. *Osteogenic potential of culture-expanded rat marrow cells as assayed in vivo with porous calcium phosphate ceramic*. Biomaterials 1991. 12:253–8. DOI: 10.1016/0142-9612(91)90209-S 108

[764] Krebsbach, PH, Kuznetsov, SA, et al. *Bone formation in vivo: comparison of osteogenesis by transplanted mouse and human marrow stromal fibroblasts*. Transplantation 1997. 63:1059–69. DOI: 10.1097/00007890-199704270-00003 108

[765] Ohgushi, H, Goldberg, VM, et al. *Repair of bone defects with marrow cells and porous ceramic. Experiments in rats.* Acta Orthop Scand 1989. 60:334–9. DOI: 10.3109/17453678909149289 108

[766] Bruder, SP, Kraus, KH, et al. *The effect of implants loaded with autologous mesenchymal stem cells on the healing of canine segmental bone defects.* J Bone Joint Surg Am 1998. 80:985–96. 108

[767] Connolly, JF, Guse, R, et al. *Autologous marrow injection for delayed unions of the tibia: a preliminary report.* J Orthop Trauma 1989. 3:276–82. DOI: 10.1097/00005131-198912000-00002 108

[768] Hernigou, P, Poignard, A, et al. *The use of percutaneous autologous bone marrow transplantation in nonunion and avascular necrosis of bone.* J Bone Joint Surg Br 2005. 87:896–902. DOI: 10.1302/0301-620X.87B7.16289 108

[769] Quarto, R, Mastrogiacomo, M, et al. *Repair of large bone defects with the use of autologous bone marrow stromal cells.* N Engl J Med 2001. 344:385–6. DOI: 10.1056/NEJM200102013440516 108

[770] Marcacci, M, Kon, E, et al. *Stem cells associated with macroporous bioceramics for long bone repair: 6- to 7-year outcome of a pilot clinical study.* Tissue Eng 2007. 13:947–55. DOI: 10.1089/ten.2006.0271 108

[771] Hesse, E, Kluge, G, et al. *Repair of a segmental long bone defect in human by implantation of a novel multiple disc graft.* Bone 2010. 46:1457–63. DOI: 10.1016/j.bone.2010.02.011 108

[772] Warnke, PH, Wiltfang, J, et al. *Man as living bioreactor: fate of an exogenously prepared customized tissue-engineered mandible.* Biomaterials 2006. 27:3163–7. DOI: 10.1016/j.biomaterials.2006.01.050 108

[773] Ueda, M, Yamada, Y, et al. *Injectable bone applied for ridge augmentation and dental implant placement: human progress study.* Implant Dent 2008. 17:82–90. DOI: 10.1097/ID.0b013e31815cd591 108

[774] Vacanti, CA, Bonassar, LJ, et al. *Replacement of an avulsed phalanx with tissue-engineered bone.* N Engl J Med 2001. 344:1511–4. DOI: 10.1056/NEJM200105173442004 108

[775] Schimming, R and Schmelzeisen, R. *Tissue-engineered bone for maxillary sinus augmentation.* J Oral Maxillofac Surg 2004. 62:724–9. DOI: 10.1016/j.joms.2004.01.009 108

[776] Horwitz, EM, Prockop, DJ, et al. *Transplantability and therapeutic effects of bone marrow-derived mesenchymal cells in children with osteogenesis imperfecta.* Nat Med 1999. 5:309–13. DOI: 10.1038/6529 108

[777] Mankin, HJ. *The reaction of articular cartilage to injury and osteoarthritis (first of two parts).* N Engl J Med 1974. 291:1285–92. DOI: 10.1056/NEJM197412192912507 109

[778] Glowacki, J and Mizuno, S. *Biomaterials in Musculoskeletal Tissue Engineering.* Current opinion in orthopaedics 2004. 12:347–354. DOI: 10.1097/01.bco.0000139326.24965.e2 109, 110

[779] Steadman, JR, Rodkey, WG, et al. *Microfracture: surgical technique and rehabilitation to treat chondral defects.* Clin Orthop Relat Res 2001. S362–9. 109

[780] Knutsen, G, Engebretsen, L, et al. *Autologous chondrocyte implantation compared with microfracture in the knee. A randomized trial.* J Bone Joint Surg Am 2004. 86-A:455–64. 109

[781] Matsusue, Y, Yamamuro, T, et al. *Arthroscopic multiple osteochondral transplantation to the chondral defect in the knee associated with anterior cruciate ligament disruption.* Arthroscopy 1993. 9:318–21. DOI: 10.1016/S0749-8063(05)80428-1 109

[782] Horas, U, Pelinkovic, D, et al. *Autologous chondrocyte implantation and osteochondral cylinder transplantation in cartilage repair of the knee joint. A prospective, comparative trial.* J Bone Joint Surg Am 2003. 85-A:185–92. 109

[783] Amiel, D, Coutts, RD, et al. *Rib perichondral grafts for the repair of full-thickness articular-cartilage defects. A morphological and biochemical study in rabbits.* J Bone Joint Surg Am 1985. 67:911–20. 109

[784] O'Driscoll, SW and Salter, RB. *The repair of major osteochondral defects in joint surfaces by neochondrogenesis with autogenous osteoperiosteal grafts stimulated by continuous passive motion. An experimental investigation in the rabbit.* Clin Orthop Relat Res 1986. 131–40. 109

[785] Roberts, S, McCall, IW, et al. *Autologous chondrocyte implantation for cartilage repair: monitoring its success by magnetic resonance imaging and histology.* Arthritis Res Ther 2003. 5:R60–73. DOI: 10.1186/ar613 109

[786] Coleman, SH, Malizia, R, et al. *Treatment of isolated articular cartilage lesions of the medial femoral condyle. A clinical nad MR comparison of autologous chondrocyte implantation vs. microfracture.* Ortop Traumatol Rehabil 2001. 3:224–6. 109

[787] Peterson, L, Brittberg, M, et al. *Autologous chondrocyte transplantation. Biomechanics and long-term durability.* Am J Sports Med 2002. 30:2–12. 109

[788] Mithofer, K, Peterson, L, et al. *Articular cartilage repair in soccer players with autologous chondrocyte transplantation: functional outcome and return to competition.* Am J Sports Med 2005. 33:1639–46. DOI: 10.1177/0363546505275647 109

[789] Minas, T, Gomoll, AH, et al. *Autologous chondrocyte implantation for joint preservation in patients with early osteoarthritis.* Clin Orthop Relat Res 2010. 468:147–57. DOI: 10.1007/s11999-009-0998-0 109

[790] Bentley, G, Biant, LC, et al. *A prospective, randomised comparison of autologous chondrocyte implantation versus mosaicplasty for osteochondral defects in the knee.* J Bone Joint Surg Br 2003. 85:223–30. DOI: 10.1302/0301-620X.85B2.13543 110

[791] Khan, IM, Gilbert, SJ, et al. *Cartilage integration: evaluation of the reasons for failure of integration during cartilage repair. A review.* Eur Cell Mater 2008. 16:26–39. 110

[792] Vinatier, C, Mrugala, D, et al. *Cartilage engineering: a crucial combination of cells, biomaterials and biofactors.* Trends Biotechnol 2009. 27:307–14. DOI: 10.1016/j.tibtech.2009.02.005 110, 118

[793] Bryant, SJ and Anseth, KS. *Hydrogel properties influence ECM production by chondrocytes photoencapsulated in poly(ethylene glycol) hydrogels.* J Biomed Mater Res 2002. 59:63–72. DOI: 10.1002/jbm.1217 110

[794] Sharma, B, Williams, CG, et al. *Designing zonal organization into tissue-engineered cartilage.* Tissue Eng 2007. 13:405–14. DOI: 10.1089/ten.2006.0068 111

[795] Woodfield, TB, Van Blitterswijk, CA, et al. *Polymer scaffolds fabricated with pore-size gradients as a model for studying the zonal organization within tissue-engineered cartilage constructs.* Tissue Eng 2005. 11:1297–311. DOI: 10.1089/ten.2005.11.1297 111

[796] Hsu, SH, Chang, SH, et al. *Evaluation of biodegradable polyesters modified by type II collagen and Arg-Gly-Asp as tissue engineering scaffolding materials for cartilage regeneration.* Artif Organs 2006. 30:42–55. DOI: 10.1111/j.1525-1594.2006.00179.x 111

[797] Yoo, HS, Lee, EA, et al. *Hyaluronic acid modified biodegradable scaffolds for cartilage tissue engineering.* Biomaterials 2005. 26:1925–33. DOI: 10.1016/j.biomaterials.2004.06.021 111

[798] Caterson, EJ, Li, WJ, et al. *Polymer/alginate amalgam for cartilage-tissue engineering.* Ann N Y Acad Sci 2002. 961:134–8. DOI: 10.1111/j.1749-6632.2002.tb03066.x 111

[799] Ameer, GA, Mahmood, TA, et al. *A biodegradable composite scaffold for cell transplantation.* J Orthop Res 2002. 20:16–9. DOI: 10.1016/S0736-0266(01)00074-2 111, 113

[800] Elisseeff, J, McIntosh, W, et al. *Controlled-release of IGF-I and TGF-beta1 in a photopolymerizing hydrogel for cartilage tissue engineering.* J Orthop Res 2001. 19:1098–104. DOI: 10.1016/S0736-0266(01)00054-7 111, 116

[801] Li, WJ, Danielson, KG, et al. *Biological response of chondrocytes cultured in three-dimensional nanofibrous poly(epsilon-caprolactone) scaffolds.* J Biomed Mater Res A 2003. 67:1105–14. DOI: 10.1002/jbm.a.10101 112

[802] Mahmood, TA, Shastri, VP, et al. *Tissue engineering of bovine articular cartilage within porous poly(ether ester) copolymer scaffolds with different structures.* Tissue Eng 2005. 11:1244–53. DOI: 10.1089/ten.2005.11.1244 112

[803] Malda, J, Woodfield, TB, et al. *The effect of PEGT/PBT scaffold architecture on the composition of tissue engineered cartilage.* Biomaterials 2005. 26:63–72. DOI: 10.1016/j.biomaterials.2004.02.046 112

[804] Martens, PJ, Bryant, SJ, et al. *Tailoring the degradation of hydrogels formed from multivinyl poly(ethylene glycol) and poly(vinyl alcohol) macromers for cartilage tissue engineering.* Biomacromolecules 2003. 4:283–92. DOI: 10.1021/bm025666v 112

[805] Temenoff, JS, Athanasiou, KA, et al. *Effect of poly(ethylene glycol) molecular weight on tensile and swelling properties of oligo(poly(ethylene glycol) fumarate) hydrogels for cartilage tissue engineering.* J Biomed Mater Res 2002. 59:429–37. DOI: 10.1002/jbm.1259 112

[806] Dadsetan, M, Szatkowski, JP, et al. *Characterization of photo-cross-linked oligo[poly(ethylene glycol) fumarate] hydrogels for cartilage tissue engineering.* Biomacromolecules 2007. 8:1702–9. DOI: 10.1021/bm070052h 112

[807] Park, H, Temenoff, JS, et al. *Injectable biodegradable hydrogel composites for rabbit marrow mesenchymal stem cell and growth factor delivery for cartilage tissue engineering.* Biomaterials 2007. 28:3217–27. DOI: 10.1016/j.biomaterials.2007.03.030 112

[808] Mueller-Rath, R, Gavenis, K, et al. *In vivo cultivation of human articular chondrocytes in a nude mouse–based contained defect organ culture model.* Biomed Mater Eng 2007. 17:357–66. 112

[809] Glowacki, J and Mizuno, S. *Collagen scaffolds for tissue engineering.* Biopolymers 2008. 89:338–44. DOI: 10.1002/bip.20871 112

[810] Yates, KE, Allemann, F, et al. *Phenotypic analysis of bovine chondrocytes cultured in 3D collagen sponges: effect of serum substitutes.* Cell Tissue Bank 2005. 6:45–54. DOI: 10.1007/s10561-005-5810-0 112

[811] Cherubino, P, Grassi, FA, et al. *Autologous chondrocyte implantation using a bilayer collagen membrane: a preliminary report.* J Orthop Surg (Hong Kong) 2003. 11:10–5. 112

[812] Behrens, P, Bitter, T, et al. *Matrix-associated autologous chondrocyte transplantation/implantation (MACT/MACI)–5-year follow-up.* Knee 2006. 13:194–202. DOI: 10.1016/j.knee.2006.02.012 112

[813] Ochi, M, Uchio, Y, et al. *Transplantation of cartilage-like tissue made by tissue engineering in the treatment of cartilage defects of the knee.* J Bone Joint Surg Br 2002. 84:571–8. DOI: 10.1302/0301-620X.84B4.11947 112

[814] Kuroda, R, Ishida, K, et al. *Treatment of a full-thickness articular cartilage defect in the femoral condyle of an athlete with autologous bone-marrow stromal cells.* Osteoarthritis Cartilage 2007. 15:226–31. DOI: 10.1016/j.joca.2006.08.008 112

[815] Aigner, J, Tegeler, J, et al. *Cartilage tissue engineering with novel nonwoven structured biomaterial based on hyaluronic acid benzyl ester.* J Biomed Mater Res 1998. 42:172–81. DOI: 10.1002/(SICI)1097-4636(199811)42:2%3C172::AID-JBM2%3E3.3.CO;2-W 112, 117

[816] Grigolo, B, Lisignoli, G, et al. *Evidence for redifferentiation of human chondrocytes grown on a hyaluronan-based biomaterial (HYAff 11): molecular, immunohistochemical and ultrastructural analysis.* Biomaterials 2002. 23:1187–95. DOI: 10.1016/S0142-9612(01)00236-8 112

[817] Radice, M, Brun, P, et al. *Hyaluronan-based biopolymers as delivery vehicles for bone-marrow-derived mesenchymal progenitors.* J Biomed Mater Res 2000. 50:101–9. DOI: 10.1002/(SICI)1097-4636(200005)50:2%3C101::AID-JBM2%3E3.0.CO;2-M 112

[818] Solchaga, LA, Temenoff, JS, et al. *Repair of osteochondral defects with hyaluronan- and polyester-based scaffolds.* Osteoarthritis Cartilage 2005. 13:297–309. DOI: 10.1016/j.joca.2004.12.016 112

[819] Smeds, KA, Pfister-Serres, A, et al. *Photocrosslinkable polysaccharides for in situ hydrogel formation.* J Biomed Mater Res 2001. 54:115–21. DOI: 10.1002/1097-4636(200101)54:1%3C115::AID-JBM14%3E3.0.CO;2-Q 112

[820] Nettles, DL, Vail, TP, et al. *Photocrosslinkable hyaluronan as a scaffold for articular cartilage repair.* Ann Biomed Eng 2004. 32:391–7. DOI: 10.1023/B:ABME.0000017552.65260.94 112

[821] Chung, C, Mesa, J, et al. *Effects of auricular chondrocyte expansion on neocartilage formation in photocrosslinked hyaluronic acid networks.* Tissue Eng 2006. 12:2665–73. DOI: 10.1089/ten.2006.12.2665 112

[822] Liu, Y, Shu, XZ, et al. *Osteochondral defect repair with autologous bone marrow-derived mesenchymal stem cells in an injectable, in situ, cross-linked synthetic extracellular matrix.* Tissue Eng 2006. 12:3405–16. DOI: 10.1089/ten.2006.12.3405 112

[823] Burdick, JA, Chung, C, et al. *Controlled degradation and mechanical behavior of photopolymerized hyaluronic acid networks.* Biomacromolecules 2005. 6:386–91. DOI: 10.1021/bm049508a 112

[824] Allemann, F, Mizuno, S, et al. *Effects of hyaluronan on engineered articular cartilage extracellular matrix gene expression in 3-dimensional collagen scaffolds.* J Biomed Mater Res 2001. 55:13–9. DOI: 10.1002/1097-4636(200104)55:1%3C13::AID-JBM20%3E3.3.CO;2-7 112

[825] Miralles, G, Baudoin, R, et al. *Sodium alginate sponges with or without sodium hyaluronate: in vitro engineering of cartilage*. J Biomed Mater Res 2001. 57:268–78. DOI: 10.1002/1097-4636(200111)57:2%3C268::AID-JBM1167%3E3.0.CO;2-L 112

[826] Yamane, S, Iwasaki, N, et al. *Feasibility of chitosan-based hyaluronic acid hybrid biomaterial for a novel scaffold in cartilage tissue engineering*. Biomaterials 2005. 26:611–9. DOI: 10.1016/j.biomaterials.2004.03.013 112

[827] Marcacci, M, Berruto, M, et al. *Articular cartilage engineering with Hyalograft C: 3-year clinical results*. Clin Orthop Relat Res 2005. 96–105. DOI: 10.1097/01.blo.0000165737.87628.5b 112

[828] Awad, HA, Wickham, MQ, et al. *Chondrogenic differentiation of adipose-derived adult stem cells in agarose, alginate, and gelatin scaffolds*. Biomaterials 2004. 25:3211–22. DOI: 10.1016/j.biomaterials.2003.10.045 113, 118

[829] Murphy, CL and Sambanis, A. *Effect of oxygen tension and alginate encapsulation on restoration of the differentiated phenotype of passaged chondrocytes*. Tissue Eng 2001. 7:791–803. DOI: 10.1089/107632701753337735

[830] Homicz, MR, Chia, SH, et al. *Human septal chondrocyte redifferentiation in alginate, polyglycolic acid scaffold, and monolayer culture*. Laryngoscope 2003. 113:25–32. DOI: 10.1097/00005537-200301000-00005 113

[831] Genes, NG, Rowley, JA, et al. *Effect of substrate mechanics on chondrocyte adhesion to modified alginate surfaces*. Arch Biochem Biophys 2004. 422:161–7. DOI: 10.1016/j.abb.2003.11.023 113

[832] Wayne, JS, McDowell, CL, et al. *In vivo response of polylactic acid-alginate scaffolds and bone marrow-derived cells for cartilage tissue engineering*. Tissue Eng 2005. 11:953–63. DOI: 10.1089/ten.2005.11.953 113

[833] Caterson, EJ, Nesti, LJ, et al. *Three-dimensional cartilage formation by bone marrow-derived cells seeded in polylactide/alginate amalgam*. J Biomed Mater Res 2001. 57:394–403. DOI: 10.1002/1097-4636(20011205)57:3%3C394::AID-JBM1182%3E3.0.CO;2-9 113

[834] Connelly, JT, Garcia, AJ, et al. *Inhibition of in vitro chondrogenesis in RGD-modified three-dimensional alginate gels*. Biomaterials 2007. 28:1071–83. DOI: 10.1016/j.biomaterials.2006.10.006 113, 116

[835] Mauck, RL, Wang, CC, et al. *The role of cell seeding density and nutrient supply for articular cartilage tissue engineering with deformational loading*. Osteoarthritis Cartilage 2003. 11:879–90. DOI: 10.1016/j.joca.2003.08.006 113

[836] Quinn, TM, Schmid, P, et al. *Proteoglycan deposition around chondrocytes in agarose culture: construction of a physical and biological interface for mechanotransduction in cartilage.* Biorheology 2002. 39:27–37. 113

[837] Mauck, RL, Soltz, MA, et al. *Functional tissue engineering of articular cartilage through dynamic loading of chondrocyte-seeded agarose gels.* J Biomech Eng 2000. 122:252–60. DOI: 10.1115/1.429656 113, 119

[838] Sechriest, VF, Miao, YJ, et al. *GAG-augmented polysaccharide hydrogel: a novel biocompatible and biodegradable material to support chondrogenesis.* J Biomed Mater Res 2000. 49:534–41. DOI: 10.1002/(SICI)1097-4636(20000315)49:4%3C534::AID-JBM12%3E3.3.CO;2-R 113

[839] Subramanian, A and Lin, HY. *Crosslinked chitosan: its physical properties and the effects of matrix stiffness on chondrocyte cell morphology and proliferation.* J Biomed Mater Res A 2005. 75:742–53. DOI: 10.1002/jbm.a.30489 113

[840] Kim, SE, Park, JH, et al. *Porous chitosan scaffold containing microspheres loaded with transforming growth factor-beta1: implications for cartilage tissue engineering.* J Control Release 2003. 91:365–74. DOI: 10.1016/S0168-3659(03)00274-8 113, 114

[841] Jeon, YH, Choi, JH, et al. *Different effects of PLGA and chitosan scaffolds on human cartilage tissue engineering.* J Craniofac Surg 2007. 18:1249–58. DOI: 10.1097/scs.0b013e3181577b55 113

[842] Wang, Y, Blasioli, DJ, et al. *Cartilage tissue engineering with silk scaffolds and human articular chondrocytes.* Biomaterials 2006. 27:4434–42. DOI: 10.1016/j.biomaterials.2006.03.050 113

[843] Wang, Y, Kim, HJ, et al. *Stem cell-based tissue engineering with silk biomaterials.* Biomaterials 2006. 27:6064–82. DOI: 10.1016/j.biomaterials.2006.07.008 113

[844] Hofmann, S, Knecht, S, et al. *Cartilage-like tissue engineering using silk scaffolds and mesenchymal stem cells.* Tissue Eng 2006. 12:2729–38. DOI: 10.1089/ten.2006.12.2729 113

[845] Svensson, A, Nicklasson, E, et al. *Bacterial cellulose as a potential scaffold for tissue engineering of cartilage.* Biomaterials 2005. 26:419–31. DOI: 10.1016/j.biomaterials.2004.02.049 113

[846] Pulkkinen, H, Tiitu, V, et al. *Cellulose sponge as a scaffold for cartilage tissue engineering.* Biomed Mater Eng 2006. 16:S29–35. 113

[847] Vinatier, C, Gauthier, O, et al. *An injectable cellulose-based hydrogel for the transfer of autologous nasal chondrocytes in articular cartilage defects.* Biotechnol Bioeng 2009. 102:1259–67. DOI: 10.1002/bit.22137 113

[848] Mosesson, MW. *Fibrinogen and fibrin structure and functions*. J Thromb Haemost 2005. 3:1894–904. DOI: 10.1111/j.1538-7836.2005.01365.x 113

[849] Sims, CD, Butler, PE, et al. *Tissue engineered neocartilage using plasma derived polymer substrates and chondrocytes*. Plast Reconstr Surg 1998. 101:1580–5. DOI: 10.1097/00006534-199805000-00022 113

[850] Passaretti, D, Silverman, RP, et al. *Cultured chondrocytes produce injectable tissue-engineered cartilage in hydrogel polymer*. Tissue Eng 2001. 7:805–15. DOI: 10.1089/107632701753337744 113

[851] Malicev, E, Radosavljevic, D, et al. *Fibrin gel improved the spatial uniformity and phenotype of human chondrocytes seeded on collagen scaffolds*. Biotechnol Bioeng 2007. 96:364–70. DOI: 10.1002/bit.21038 113

[852] Eyrich, D, Brandl, F, et al. *Long-term stable fibrin gels for cartilage engineering*. Biomaterials 2007. 28:55–65. DOI: 10.1016/j.biomaterials.2006.08.027 113

[853] Kim, MK, Choi, SW, et al. *Autologous chondrocyte implantation in the knee using fibrin*. Knee Surg Sports Traumatol Arthrosc 2010. 18:528–34. DOI: 10.1007/s00167-009-0905-y 114

[854] Saraf, A and Mikos, AG. *Gene delivery strategies for cartilage tissue engineering*. Adv Drug Deliv Rev 2006. 58:592–603. DOI: 10.1016/j.addr.2006.03.005 114

[855] Trippel, SB, Ghivizzani, SC, et al. *Gene-based approaches for the repair of articular cartilage*. Gene Ther 2004. 11:351–9. DOI: 10.1038/sj.gt.3302201 114

[856] Grande, DA, Mason, J, et al. *Stem cells as platforms for delivery of genes to enhance cartilage repair*. J Bone Joint Surg Am 2003. 85-A Suppl 2:111–6. 114, 115

[857] Guo, T, Zhao, J, et al. *Porous chitosan-gelatin scaffold containing plasmid DNA encoding transforming growth factor-beta1 for chondrocytes proliferation*. Biomaterials 2006. 27:1095–103. DOI: 10.1016/j.biomaterials.2005.08.015 114

[858] Capito, RM and Spector, M. *Collagen scaffolds for nonviral IGF-1 gene delivery in articular cartilage tissue engineering*. Gene Ther 2007. 14:721–32. DOI: 10.1038/sj.gt.3302918 114, 116

[859] Jin, XB, Sun, YS, et al. *Tissue engineered cartilage from hTGF beta2 transduced human adipose derived stem cells seeded in PLGA/alginate compound in vitro and in vivo*. J Biomed Mater Res A 2008. 86:1077–87. DOI: 10.1002/jbm.a.31656 114

[860] Hwang, NS, Kim, MS, et al. *Effects of three-dimensional culture and growth factors on the chondrogenic differentiation of murine embryonic stem cells*. Stem Cells 2006. 24:284–91. DOI: 10.1634/stemcells.2005-0024 114, 118

[861] Iwasaki, M, Nakata, K, et al. *Transforming growth factor-beta 1 stimulates chondrogenesis and inhibits osteogenesis in high density culture of periosteum-derived cells.* Endocrinology 1993. 132:1603–8. DOI: 10.1210/en.132.4.1603 114

[862] Rosier, RN, O'Keefe, RJ, et al. *Transforming growth factor beta: an autocrine regulator of chondrocytes.* Connect Tissue Res 1989. 20:295–301. DOI: 10.3109/03008208909023900 114

[863] Vivien, D, Galera, P, et al. *Differential effects of transforming growth factor-beta and epidermal growth factor on the cell cycle of cultured rabbit articular chondrocytes.* J Cell Physiol 1990. 143:534–45. DOI: 10.1002/jcp.1041430319 114

[864] Fan, H, Hu, Y, et al. *Porous gelatin-chondroitin-hyaluronate tri-copolymer scaffold containing microspheres loaded with TGF-beta1 induces differentiation of mesenchymal stem cells in vivo for enhancing cartilage repair.* J Biomed Mater Res A 2006. 77:785–94. DOI: 10.1002/jbm.a.30647 114, 118, 120

[865] Na, K, Park, JH, et al. *Delivery of dexamethasone, ascorbate, and growth factor (TGF beta-3) in thermo-reversible hydrogel constructs embedded with rabbit chondrocytes.* Biomaterials 2006. 27:5951–7. DOI: 10.1016/j.biomaterials.2006.08.012 114

[866] Lee, JE, Kim, SE, et al. *Effects of a chitosan scaffold containing TGF-beta1 encapsulated chitosan microspheres on in vitro chondrocyte culture.* Artif Organs 2004. 28:829–39. DOI: 10.1111/j.1525-1594.2004.00020.x 114

[867] Sharma, B, Williams, CG, et al. *In vivo chondrogenesis of mesenchymal stem cells in a photopoly-merized hydrogel.* Plast Reconstr Surg 2007. 119:112–20. DOI: 10.1097/01.prs.0000236896.22479.52 114, 121

[868] Bilgen, B, Ren, Y, et al. *CD14-negative isolation enhances chondrogenesis in synovial fibroblasts.* Tissue Eng Part A 2009. 15:3261–70. DOI: 10.1089/ten.tea.2008.0273 114, 118

[869] Huang, Q, Goh, JC, et al. *In vivo mesenchymal cell recruitment by a scaffold loaded with transforming growth factor beta1 and the potential for in situ chondrogenesis.* Tissue Eng 2002. 8:469–82. DOI: 10.1089/107632702760184727 114

[870] Tong, Jc and Sao, yl. *Novel scaffold containing transforming growth factor-beta 1 DNA for cartilage tissue engineering.* J Bioactive Compatible Polymers 2007. 22:232–244. DOI: 10.1177/0883911507076460 115

[871] van Beuningen, HM, Glansbeek, HL, et al. *Differential effects of local application of BMP-2 or TGF-beta 1 on both articular cartilage composition and osteophyte formation.* Osteoarthritis Cartilage 1998. 6:306–17. DOI: 10.1053/joca.1998.0129 115

[872] Sekiya, I, Larson, BL, et al. *Comparison of effect of BMP-2, –4, and –6 on in vitro cartilage formation of human adult stem cells from bone marrow stroma.* Cell Tissue Res 2005. 320:269–76. DOI: 10.1007/s00441-004-1075-3 115

[873] Grunder, T, Gaissmaier, C, et al. *Bone morphogenetic protein (BMP)-2 enhances the expression of type II collagen and aggrecan in chondrocytes embedded in alginate beads.* Osteoarthritis Cartilage 2004. 12:559–67. DOI: 10.1016/j.joca.2004.04.001 115

[874] Park, Y, Sugimoto, M, et al. *BMP-2 induces the expression of chondrocyte-specific genes in bovine synovium-derived progenitor cells cultured in three-dimensional alginate hydrogel.* Osteoarthritis Cartilage 2005. 13:527–36. DOI: 10.1016/j.joca.2005.02.006 115, 118

[875] Kuh, SU, Zhu, Y, et al. *Can TGF-beta1 and rhBMP-2 act in synergy to transform bone marrow stem cells to discogenic-type cells?* Acta Neurochir (Wien) 2008. 150:1073–9; discussion 1079. DOI: 10.1007/s00701-008-0029-z 115

[876] Kaps, C, Bramlage, C, et al. *Bone morphogenetic proteins promote cartilage differentiation and protect engineered artificial cartilage from fibroblast invasion and destruction.* Arthritis Rheum 2002. 46:149–62. DOI: 10.1002/1529-0131(200201)46:1%3C149::AID-ART10058%3E3.0.CO;2-W 115

[877] Hicks, DL, Sage, AB, et al. *Effect of bone morphogenetic proteins 2 and 7 on septal chondrocytes in alginate.* Otolaryngol Head Neck Surg 2007. 136:373–9. DOI: 10.1016/j.otohns.2006.10.040 115

[878] Kuo, AC, Rodrigo, JJ, et al. *Microfracture and bone morphogenetic protein 7 (BMP-7) synergistically stimulate articular cartilage repair.* Osteoarthritis Cartilage 2006. 14:1126–35. DOI: 10.1016/j.joca.2006.04.004 115

[879] Di Cesare, PE, Frenkel, SR, et al. *Regional gene therapy for full-thickness articular cartilage lesions using naked DNA with a collagen matrix.* J Orthop Res 2006. 24:1118–27. DOI: 10.1002/jor.20143 115

[880] Kuroda, R, Usas, A, et al. *Cartilage repair using bone morphogenetic protein 4 and muscle-derived stem cells.* Arthritis Rheum 2006. 54:433–42. DOI: 10.1002/art.21632 115

[881] Stewart, AA, Byron, CR, et al. *Effect of fibroblast growth factor-2 on equine mesenchymal stem cell monolayer expansion and chondrogenesis.* Am J Vet Res 2007. 68:941–5. DOI: 10.2460/ajvr.68.9.941 115

[882] Martin, I, Suetterlin, R, et al. *Enhanced cartilage tissue engineering by sequential exposure of chondrocytes to FGF-2 during 2D expansion and BMP-2 during 3D cultivation.* J Cell Biochem 2001. 83:121–8. DOI: 10.1002/jcb.1203 115

[883] Akaogi, H, Akmoto, T, et al. *Basic fibroblast growth factor supports in vitro chondrogenesis of bone marrow-derived mesenchymal stem cells from patients with osteoarthritis.* Mat Sci Eng 2004. C24:403–406. DOI: 10.1016/j.msec.2003.12.017 115

[884] Jansen, EJ, Emans, PJ, et al. *Human periosteum-derived cells from elderly patients as a source for cartilage tissue engineering?* J Tissue Eng Regen Med 2008. 2:331–9. DOI: 10.1002/term.100 115

[885] Veilleux, N and Spector, M. *Effects of FGF-2 and IGF-1 on adult canine articular chondrocytes in type II collagen-glycosaminoglycan scaffolds in vitro.* Osteoarthritis Cartilage 2005. 13:278–86. DOI: 10.1016/j.joca.2004.12.013 115

[886] Fujisato, T, Sajiki, T, et al. *Effect of basic fibroblast growth factor on cartilage regeneration in chondrocyte-seeded collagen sponge scaffold.* Biomaterials 1996. 17:155–62. DOI: 10.1016/0142-9612(96)85760-7 115

[887] Ishii, I, Mizuta, H, et al. *Healing of full-thickness defects of the articular cartilage in rabbits using fibroblast growth factor-2 and a fibrin sealant.* J Bone Joint Surg Br 2007. 89:693–700. DOI: 10.1302/0301-620X.89B5.18450 115

[888] Huang, X, Yang, D, et al. *Osteochondral repair using the combination of fibroblast growth factor and amorphous calcium phosphate/poly(L-lactic acid) hybrid materials.* Biomaterials 2007. 28:3091–100. DOI: 10.1016/j.biomaterials.2007.03.017 115

[889] Davies, LC, Blain, EJ, et al. *The potential of IGF-1 and TGFbeta1 for promoting "adult" articular cartilage repair: an in vitro study.* Tissue Eng Part A 2008. 14:1251–61. DOI: 10.1089/ten.tea.2007.0211 115

[890] Uebersax, L, Merkle, HP, et al. *Insulin-like growth factor I releasing silk fibroin scaffolds induce chondrogenic differentiation of human mesenchymal stem cells.* J Control Release 2008. 127:12–21. DOI: 10.1016/j.jconrel.2007.11.006 116

[891] Madry, H, Padera, R, et al. *Gene transfer of a human insulin-like growth factor I cDNA enhances tissue engineering of cartilage.* Hum Gene Ther 2002. 13:1621–30. DOI: 10.1089/10430340260201716 116

[892] Mauck, RL, Nicoll, SB, et al. *Synergistic action of growth factors and dynamic loading for articular cartilage tissue engineering.* Tissue Eng 2003. 9:597–611. DOI: 10.1089/107632703768247304 116

[893] Yaeger, PC, Masi, TL, et al. *Synergistic action of transforming growth factor-beta and insulin-like growth factor-I induces expression of type II collagen and aggrecan genes in adult human articular chondrocytes.* Exp Cell Res 1997. 237:318–25. DOI: 10.1006/excr.1997.3781 116

[894] Holland, TA, Bodde, EW, et al. *Degradable hydrogel scaffolds for in vivo delivery of single and dual growth factors in cartilage repair.* Osteoarthritis Cartilage 2007. 15:187–97. DOI: 10.1016/j.joca.2006.07.006 116

[895] Bryant, SJ, Nicodemus, GD, et al. *Designing 3D photopolymer hydrogels to regulate biomechanical cues and tissue growth for cartilage tissue engineering.* Pharm Res 2008. 25:2379–86. DOI: 10.1007/s11095-008-9619-y 116

[896] Hwang, NS, Varghese, S, et al. *Chondrogenic differentiation of human embryonic stem cell-derived cells in arginine-glycine-aspartate-modified hydrogels.* Tissue Eng 2006. 12:2695–706. DOI: 10.1089/ten.2006.12.2695 116, 118

[897] Tigli, RS and Gumusderelioglu, M. *Evaluation of RGD- or EGF-immobilized chitosan scaffolds for chondrogenic activity.* Int J Biol Macromol 2008. 43:121–8. DOI: 10.1016/j.ijbiomac.2008.04.003 116

[898] Schnabel, M, Marlovits, S, et al. *Dedifferentiation-associated changes in morphology and gene expression in primary human articular chondrocytes in cell culture.* Osteoarthritis Cartilage 2002. 10:62–70. DOI: 10.1053/joca.2001.0482 116

[899] Darling, EM and Athanasiou, KA. *Rapid phenotypic changes in passaged articular chondrocyte subpopulations.* J Orthop Res 2005. 23:425–32. DOI: 10.1016/j.orthres.2004.08.008 116

[900] Van Osch, GJ, Mandl, EW, et al. *Considerations on the use of ear chondrocytes as donor chondrocytes for cartilage tissue engineering.* Biorheology 2004. 41:411–21. 117

[901] Panossian, A, Ashiku, S, et al. *Effects of cell concentration and growth period on articular and ear chondrocyte transplants for tissue engineering.* Plast Reconstr Surg 2001. 108:392–402. DOI: 10.1097/00006534-200108000-00018 117

[902] Kafienah, W, Jakob, M, et al. *Three-dimensional tissue engineering of hyaline cartilage: comparison of adult nasal and articular chondrocytes.* Tissue Eng 2002. 8:817–26. DOI: 10.1089/10763270260424178 117

[903] Naumann, A, Dennis, JE, et al. *Tissue engineering of autologous cartilage grafts in three-dimensional in vitro macroaggregate culture system.* Tissue Eng 2004. 10:1695–706. DOI: 10.1089/ten.2004.10.1695 117

[904] Shikani, AH, Fink, DJ, et al. *Propagation of human nasal chondrocytes in microcarrier spinner culture.* Am J Rhinol 2004. 18:105–12.

[905] Miot, S, Woodfield, T, et al. *Effects of scaffold composition and architecture on human nasal chondrocyte redifferentiation and cartilaginous matrix deposition.* Biomaterials 2005. 26:2479–89. DOI: 10.1016/j.biomaterials.2004.06.048

[906] Vinatier, C, Magne, D, et al. *Engineering cartilage with human nasal chondrocytes and a silanized hydroxypropyl methylcellulose hydrogel.* J Biomed Mater Res A 2007. 80:66–74. DOI: 10.1002/jbm.a.30867 117

[907] Richmon, JD, Sage, AB, et al. *Effect of growth factors on cell proliferation, matrix deposition, and morphology of human nasal septal chondrocytes cultured in monolayer.* Laryngoscope 2005. 115:1553–60. DOI: 10.1097/01.MLG.0000175541.31131.A5 117

[908] van Osch, GJ, Marijnissen, WJ, et al. *The potency of culture-expanded nasal septum chondrocytes for tissue engineering of cartilage.* Am J Rhinol 2001. 15:187–92. DOI: 10.2500/105065801779954166 117

[909] Isogai, N, Kusuhara, H, et al. *Comparison of different chondrocytes for use in tissue engineering of cartilage model structures.* Tissue Eng 2006. 12:691–703. DOI: 10.1089/ten.2006.12.691 117

[910] Tay, AG, Farhadi, J, et al. *Cell yield, proliferation, and postexpansion differentiation capacity of human ear, nasal, and rib chondrocytes.* Tissue Eng 2004. 10:762–70. DOI: 10.1089/1076327041348572 117

[911] Terada, S, Fuchs, JR, et al. *In vitro cartilage regeneration from proliferated adult elastic chondrocytes.* Ann Plast Surg 2005. 55:196–201. DOI: 10.1097/01.sap.0000164388.33965.4e 117

[912] Acosta, CA, Izal, I, et al. *Gene expression and proliferation analysis in young, aged, and osteoarthritic sheep chondrocytes effect of growth factor treatment.* J Orthop Res 2006. 24:2087–94. DOI: 10.1002/jor.20245 117

[913] Dorotka, R, Bindreiter, U, et al. *Behavior of human articular chondrocytes derived from nonarthritic and osteoarthritic cartilage in a collagen matrix.* Tissue Eng 2005. 11:877–86. DOI: 10.1089/ten.2005.11.877 117

[914] Hsieh-Bonassera, ND, Wu, I, et al. *Expansion and redifferentiation of chondrocytes from osteoarthritic cartilage: cells for human cartilage tissue engineering.* Tissue Eng Part A 2009. 15:3513–23. DOI: 10.1089/ten.tea.2008.0628 117

[915] Jakob, M, Demarteau, O, et al. *Specific growth factors during the expansion and redifferentiation of adult human articular chondrocytes enhance chondrogenesis and cartilaginous tissue formation in vitro.* J Cell Biochem 2001. 81:368–77. DOI: 10.1002/1097-4644(20010501)81:2%3C368::AID-JCB1051%3E3.0.CO;2-J 117

[916] Carossino, AM, Recenti, R, et al. *Methodological models for in vitro amplification and maintenance of human articular chondrocytes from elderly patients.* Biogerontology 2007. 8:483–98. DOI: 10.1007/s10522-007-9088-4 117

[917] Mesa, JM, Zaporojan, V, et al. *Tissue engineering cartilage with aged articular chondrocytes in vivo*. Plast Reconstr Surg 2006. 118:41–9; discussion 50–3. DOI: 10.1097/01.prs.0000231929.37736.28 117

[918] De Bari, C, Dell'Accio, F, et al. *Human periosteum-derived cells maintain phenotypic stability and chondrogenic potential throughout expansion regardless of donor age*. Arthritis Rheum 2001. 44:85–95. DOI: 10.1002/1529-0131(200101)44:1%3C85::AID-ANR12%3E3.0.CO;2-6 117

[919] Marlovits, S, Tichy, B, et al. *Chondrogenesis of aged human articular cartilage in a scaffold-free bioreactor*. Tissue Eng 2003. 9:1215–26. DOI: 10.1089/10763270360728125 117

[920] Dragoo, JL, Samimi, B, et al. *Tissue-engineered cartilage and bone using stem cells from human infrapatellar fat pads*. J Bone Joint Surg Br 2003. 85:740–7. 117

[921] Wickham, MQ, Erickson, GR, et al. *Multipotent stromal cells derived from the infrapatellar fat pad of the knee*. Clin Orthop Relat Res 2003. 196–212. DOI: 10.1097/01.blo.0000072467.53786.ca 117

[922] Noel, D, Djouad, F, et al. *Multipotent mesenchymal stromal cells and immune tolerance*. Leuk Lymphoma 2007. 48:1283–9. DOI: 10.1080/10428190701361869 118

[923] Zhou, S, Yates, KE, et al. *Demineralized bone promotes chondrocyte or osteoblast differentiation of human marrow stromal cells cultured in collagen sponges*. Cell Tissue Bank 2005. 6:33–44. DOI: 10.1007/s10561-005-4253-y 118

[924] Huang, CY, Reuben, PM, et al. *Chondrogenesis of human bone marrow-derived mesenchymal stem cells in agarose culture*. Anat Rec A Discov Mol Cell Evol Biol 2004. 278:428–36. DOI: 10.1002/ar.a.20010

[925] Williams, CG, Kim, TK, et al. *In vitro chondrogenesis of bone marrow-derived mesenchymal stem cells in a photopolymerizing hydrogel*. Tissue Eng 2003. 9:679–88. DOI: 10.1089/107632703768247377 118

[926] Tsuchiya, K, Chen, GP, et al. *The effect of Coculture of chondrocytes with mesenchymal stem cells on their cartilaginous phenotype in vitro*. Mat Sci Eng C 2004. 23:391–396. DOI: 10.1016/j.msec.2003.12.014 118

[927] Huang, JI, Zuk, PA, et al. *Chondrogenic potential of multipotential cells from human adipose tissue*. Plast Reconstr Surg 2004. 113:585–94. DOI: 10.1097/01.PRS.0000101063.27008.E1 118

[928] Ogawa, R, Mizuno, H, et al. *Osteogenic and chondrogenic differentiation by adipose-derived stem cells harvested from GFP transgenic mice*. Biochem Biophys Res Commun 2004. 313:871–7. DOI: 10.1016/j.bbrc.2003.12.017 118

[929] Zuk, PA, Zhu, M, et al. *Human adipose tissue is a source of multipotent stem cells.* Mol Biol Cell 2002. 13:4279–95. DOI: 10.1091/mbc.E02-02-0105 118

[930] Chiou, M, Xu, Y, et al. *Mitogenic and chondrogenic effects of fibroblast growth factor-2 in adipose-derived mesenchymal cells.* Biochem Biophys Res Commun 2006. 343:644–52. DOI: 10.1016/j.bbrc.2006.02.171 118

[931] Masuoka, K, Asazuma, T, et al. *Tissue engineering of articular cartilage with autologous cultured adipose tissue-derived stromal cells using atelocollagen honeycomb-shaped scaffold with a membrane sealing in rabbits.* J Biomed Mater Res B Appl Biomater 2006. 79:25–34. DOI: 10.1002/jbm.b.30507 118

[932] Winter, A, Breit, S, et al. *Cartilage-like gene expression in differentiated human stem cell spheroids: a comparison of bone marrow-derived and adipose tissue-derived stromal cells.* Arthritis Rheum 2003. 48:418–29. DOI: 10.1002/art.10767 118

[933] Adachi, N, Sato, K, et al. *Muscle derived, cell based ex vivo gene therapy for treatment of full thickness articular cartilage defects.* J Rheumatol 2002. 29:1920–30. 118

[934] Ming, P, Fan, H, et al. *Engineering of functional cartilage tissue using stem cells from synovial lining: a preliminary study.* Clin Orthop Rel Res 2008. 466:1880–1889. DOI: 10.1007/s11999-008-0316-2 118

[935] Sakaguchi, Y, Sekiya, I, et al. *Comparison of human stem cells derived from various mesenchymal tissues: superiority of synovium as a cell source.* Arthritis Rheum 2005. 52:2521–9. DOI: 10.1002/art.21212 118

[936] Kramer, J, Hegert, C, et al. *Embryonic stem cell-derived chondrogenic differentiation in vitro: activation by BMP-2 and BMP-4.* Mech Dev 2000. 92:193–205. DOI: 10.1016/S0925-4773(99)00339-1 118

[937] Vats, A, Bielby, RC, et al. *Chondrogenic differentiation of human embryonic stem cells: the effect of the micro-environment.* Tissue Eng 2006. 12:1687–97. DOI: 10.1089/ten.2006.12.1687 118

[938] Koay, EJ, Hoben, GM, et al. *Tissue engineering with chondrogenically differentiated human embryonic stem cells.* Stem Cells 2007. 25:2183–90. DOI: 10.1634/stemcells.2007-0105 118

[939] Yates, KE, Forbes, RL, et al. *New chondrocyte genes discovered by representational difference analysis of chondroinduced human fibroblasts.* Cells Tissues Organs 2004. 176:41–53. DOI: 10.1159/000075026 118

[940] French, MM, Rose, S, et al. *Chondrogenic differentiation of adult dermal fibroblasts.* Ann Biomed Eng 2004. 32:50–6. DOI: 10.1023/B:ABME.0000007790.65773.e0 118

[941] Lee, KH, Song, SU, et al. *Regeneration of hyaline cartilage by cell-mediated gene therapy using transforming growth factor beta 1-producing fibroblasts.* Hum Gene Ther 2001. 12:1805–13. DOI: 10.1089/104303401750476294 118

[942] Vunjak-Novakovic, G, Obradovic, B, et al. *Dynamic cell seeding of polymer scaffolds for cartilage tissue engineering.* Biotechnol Prog 1998. 14:193–202. DOI: 10.1021/bp970120j 119

[943] Bueno, EM, Bilgen, B, et al. *Wavy-walled bioreactor supports increased cell proliferation and matrix deposition in engineered cartilage constructs.* Tissue Eng 2005. 11:1699–709. DOI: 10.1089/ten.2005.11.1699

[944] Bueno, EM, Bilgen, B, et al. *Hydrodynamic parameters modulate biochemical, histological, and mechanical properties of engineered cartilage.* Tissue Eng Part A 2009. 15:773–85. DOI: 10.1089/ten.tea.2008.0081

[945] Freed, LE and Vunjak-Novakovic, G. *Cultivation of cell-polymer tissue constructs in simulated microgravity.* Biotechnol Bioeng 1995. 46:306–13. DOI: 10.1002/bit.260460403

[946] Mizuno, S, Tateishi, T, et al. *Hydrostatic fluid pressure enhances matrix synthesis and accumulation by bovine chondrocytes in three-dimensional culture.* J Cell Physiol 2002. 193:319–27. DOI: 10.1002/jcp.10180 119

[947] Seidel, JO, Pei, M, et al. *Long-term culture of tissue engineered cartilage in a perfused chamber with mechanical stimulation.* Biorheology 2004. 41:445–58. 119

[948] Carver, SE and Heath, CA. *Semi-continuous perfusion system for delivering intermittent physiological pressure to regenerating cartilage.* Tissue Eng 1999. 5:1–11. DOI: 10.1089/ten.1999.5.1 119

[949] Gavenis, K, Kremer, A, et al. *Effects of cyclic hydrostatic pressure on the metabolism of human osteoarthritic chondrocytes cultivated in a collagen gel.* Artif Organs 2007. 31:91–8. DOI: 10.1111/j.1525-1594.2007.00347.x

[950] Scherer, K, Schunke, M, et al. *The influence of oxygen and hydrostatic pressure on articular chondrocytes and adherent bone marrow cells in vitro.* Biorheology 2004. 41:323–33. 119

[951] Miyanishi, K, Trindade, MC, et al. *Effects of hydrostatic pressure and transforming growth factor-beta 3 on adult human mesenchymal stem cell chondrogenesis in vitro.* Tissue Eng 2006. 12:1419–28. DOI: 10.1089/ten.2006.12.1419

[952] Hu, JC and Athanasiou, KA. *The effects of intermittent hydrostatic pressure on self-assembled articular cartilage constructs.* Tissue Eng 2006. 12:1337–44. DOI: 10.1089/ten.2006.12.1337 119

[953] Hansen, U, Schunke, M, et al. *Combination of reduced oxygen tension and intermittent hydrostatic pressure: a useful tool in articular cartilage tissue engineering.* J Biomech 2001. 34:941–9. DOI: 10.1016/S0021-9290(01)00050-1 119

[954] Stoddart, MJ, Ettinger, L, et al. *Enhanced matrix synthesis in de novo, scaffold free cartilage-like tissue subjected to compression and shear.* Biotechnol Bioeng 2006. 95:1043–51. DOI: 10.1002/bit.21052 119

[955] Davisson, T, Kunig, S, et al. *Static and dynamic compression modulate matrix metabolism in tissue engineered cartilage.* J Orthop Res 2002. 20:842–8. DOI: 10.1016/S0736-0266(01)00160-7 119

[956] Lee, CR, Grodzinsky, AJ, et al. *Biosynthetic response of passaged chondrocytes in a type II collagen scaffold to mechanical compression.* J Biomed Mater Res A 2003. 64:560–9. DOI: 10.1002/jbm.a.10443 119

[957] De Croos, JN, Dhaliwal, SS, et al. *Cyclic compressive mechanical stimulation induces sequential catabolic and anabolic gene changes in chondrocytes resulting in increased extracellular matrix accumulation.* Matrix Biol 2006. 25:323–31. DOI: 10.1016/j.matbio.2006.03.005 119

[958] Rajpurohit, R, Koch, CJ, et al. *Adaptation of chondrocytes to low oxygen tension: relationship between hypoxia and cellular metabolism.* J Cell Physiol 1996. 168:424–32. DOI: 10.1002/(SICI)1097-4652(199608)168:2%3C424::AID-JCP21%3E3.0.CO;2-1 119

[959] McNulty, AL, Stabler, TV, et al. *Dehydroascorbate transport in human chondrocytes is regulated by hypoxia and is a physiologically relevant source of ascorbic acid in the joint.* Arthritis Rheum 2005. 52:2676–85. DOI: 10.1002/art.21254 119

[960] Saini, S and Wick, TM. *Effect of low oxygen tension on tissue-engineered cartilage construct development in the concentric cylinder bioreactor.* Tissue Eng 2004. 10:825–32. DOI: 10.1089/1076327041348545 119

[961] Kurz, B, Domm, C, et al. *Tissue engineering of articular cartilage under the influence of collagen I/III membranes and low oxygen tension.* Tissue Eng 2004. 10:1277–86. DOI: 10.1089/ten.2004.10.1277

[962] Wernike, E, Li, Z, et al. *Effect of reduced oxygen tension and long-term mechanical stimulation on chondrocyte-polymer constructs.* Cell Tissue Res 2008. 331:473–83. DOI: 10.1007/s00441-007-0500-9

[963] Mizuno, S and Glowacki, J. *Low oxygen tension enhances chondroinduction by demineralized bone matrix in human dermal fibroblasts in vitro.* Cells Tissues Organs 2005. 180:151–8. DOI: 10.1159/000088243 119

[964] Obradovic, B, Carrier, RL, et al. *Gas exchange is essential for bioreactor cultivation of tissue engineered cartilage.* Biotechnol Bioeng 1999. 63:197–205.
DOI: 10.1002/(SICI)1097-0290(19990420)63:2%3C197::AID-BIT8%3E3.0.CO;2-2
119

[965] Grimshaw, MJ and Mason, RM. *Modulation of bovine articular chondrocyte gene expression in vitro by oxygen tension.* Osteoarthritis Cartilage 2001. 9:357–64.
DOI: 10.1053/joca.2000.0396 119

[966] Bilgen, B, Uygun, K, et al. *Tissue growth modeling in a wavy-walled bioreactor.* Tissue Eng Part A 2009. 15:761–71. DOI: 10.1089/ten.tea.2008.0078 119

[967] Freed, LE, Marquis, JC, et al. *Kinetics of chondrocyte growth in cell-polymer implants.* Biotechnol Bioeng 1994. 43:597–604. DOI: 10.1002/bit.260430709 119

[968] Wilson, CG, Bonassar, LJ, et al. *Modeling the dynamic composition of engineered cartilage.* Arch Biochem Biophys 2002. 408:246–54. DOI: 10.1016/S0003-9861(02)00562-3

[969] Kino-Oka, M, Maeda, Y, et al. *A kinetic modeling of chondrocyte culture for manufacture of tissue-engineered cartilage.* J Biosci Bioeng 2005. 99:197–207. DOI: 10.1263/jbb.99.197 119

[970] Saha, AK, Mazumdar, J, et al. *Prediction of growth factor effects on engineered cartilage composition using deterministic and stochastic modeling.* Ann Biomed Eng 2004. 32:871–9.
DOI: 10.1023/B:ABME.0000030262.82626.9c 119

[971] Vunjak-Novakovic, G, Martin, I, et al. *Bioreactor cultivation conditions modulate the composition and mechanical properties of tissue-engineered cartilage.* J Orthop Res 1999. 17:130–8.
DOI: 10.1002/jor.1100170119 119

[972] Galban, CJ and Locke, BR. *Analysis of cell growth kinetics and substrate diffusion in a polymer scaffold.* Biotechnol Bioeng 1999. 65:121–32.
DOI: 10.1002/(SICI)1097-0290(19991020)65:2%3C121::AID-BIT1%3E3.0.CO;2-6

[973] Galbusera, F, Cioffi, M, et al. *Computational modeling of combined cell population dynamics and oxygen transport in engineered tissue subject to interstitial perfusion.* Comput Methods Biomech Biomed Engin 2007. 10:279–87. DOI: 10.1080/10255840701318404 119

[974] Mauck, RL, Hung, CT, et al. *Modeling of neutral solute transport in a dynamically loaded porous permeable gel: implications for articular cartilage biosynthesis and tissue engineering.* J Biomech Eng 2003. 125:602–14. DOI: 10.1115/1.1611512 120

[975] Bueno, EM, Laevsky, G, et al. *Enhancing cell seeding of scaffolds in tissue engineering through manipulation of hydrodynamic parameters.* J Biotechnol 2007. 129:516–31.
DOI: 10.1016/j.jbiotec.2007.01.005 120

[976] Freed, LE, Marquis, JC, et al. *Composition of cell-polymer cartilage implants*. Biotechnol Bioeng 1994. 43:605–14. DOI: 10.1002/bit.260430710 120

[977] Hung, CT, Lima, EG, et al. *Anatomically shaped osteochondral constructs for articular cartilage repair*. J Biomech 2003. 36:1853–64. DOI: 10.1016/S0021-9290(03)00213-6 120

[978] Demarteau, O, Wendt, D, et al. *Dynamic compression of cartilage constructs engineered from expanded human articular chondrocytes*. Biochem Biophys Res Commun 2003. 310:580–8. DOI: 10.1016/j.bbrc.2003.09.099 120

[979] Kreuz, PC, Muller, S, et al. *Treatment of focal degenerative cartilage defects with polymer-based autologous chondrocyte grafts: four-year clinical results*. Arthritis Res Ther 2009. 11:R33. DOI: 10.1186/ar2638 120

[980] Nehrer, S, Dorotka, R, et al. *Treatment of full-thickness chondral defects with hyalograft C in the knee: a prospective clinical case series with 2 to 7 years' follow-up*. Am J Sports Med 2009. 37 Suppl 1:81S-87S. DOI: 10.1177/0363546509350704 120

[981] Crawford, DC, Heveran, CM, et al. *An autologous cartilage tissue implant NeoCart for treatment of grade III chondral injury to the distal femur: prospective clinical safety trial at 2 years*. Am J Sports Med 2009. 37:1334–43. DOI: 10.1177/0363546509333011 120

[982] Han, SH, Kim, YH, et al. *Histological and biomechanical properties of regenerated articular cartilage using chondrogenic bone marrow stromal cells with a PLGA scaffold in vivo*. J Biomed Mater Res A 2008. 87:850–61. DOI: 10.1002/jbm.a.31828 120

[983] Liu, L, Chen, RL, et al. *[A study on the chondrogenesis of the compound of alginate gelatin and bone marrow stromal cells in vivo]*. Hua Xi Kou Qiang Yi Xue Za Zhi 2005. 23:60–2. 121

[984] Jin, M, Frank, EH, et al. *Tissue shear deformation stimulates proteoglycan and protein biosynthesis in bovine cartilage explants*. Arch Biochem Biophys 2001. 395:41–8. DOI: 10.1006/abbi.2001.2543 120, 121

[985] Erickson, GR, Gimble, JM, et al. *Chondrogenic potential of adipose tissue-derived stromal cells in vitro and in vivo*. Biochem Biophys Res Commun 2002. 290:763–9. DOI: 10.1006/bbrc.2001.6270 121

[986] Petersen, JP, Ueblacker, P, et al. *Long term results after implantation of tissue engineered cartilage for the treatment of osteochondral lesions in a minipig model*. J Mater Sci Mater Med 2008. 19:2029–38. DOI: 10.1007/s10856-007-3291-3 121

[987] Fecek, C, Yao, D, et al. *Chondrogenic derivatives of embryonic stem cells seeded into 3D polycaprolactone scaffolds generated cartilage tissue in vivo*. Tissue Eng Part A 2008. 14:1403–13. DOI: 10.1089/ten.tea.2007.0293 121

[988] Munirah, S, Kim, SH, et al. *The use of fibrin and poly(lactic-co-glycolic acid) hybrid scaffold for articular cartilage tissue engineering: an in vivo analysis.* Eur Cell Mater 2008. 15:41–52. 121

[989] Yan, J, Qi, N, et al. *Rabbit articular chondrocytes seeded on collagen-chitosan-GAG scaffold for cartilage tissue engineering in vivo.* Artif Cells Blood Substit Immobil Biotechnol 2007. 35:333–44. DOI: 10.1080/10731190701460200

[990] Liao, E, Yaszemski, M, et al. *Tissue-engineered cartilage constructs using composite hyaluronic acid/collagen I hydrogels and designed poly(propylene fumarate) scaffolds.* Tissue Eng 2007. 13:537–50. DOI: 10.1089/ten.2006.0117

[991] Jin, CZ, Park, SR, et al. *In vivo cartilage tissue engineering using a cell-derived extracellular matrix scaffold.* Artif Organs 2007. 31:183–92. DOI: 10.1111/j.1525-1594.2007.00363.x 121

[992] Pereira, RC, Scaranari, M, et al. *Novel injectable gel (system) as a vehicle for human articular chondrocytes in cartilage tissue regeneration.* J Tissue Eng Regen Med 2009. 3:97–106. DOI: 10.1002/term.145 121

[993] Shangkai, C, Naohide, T, et al. *Transplantation of allogeneic chondrocytes cultured in fibroin sponge and stirring chamber to promote cartilage regeneration.* Tissue Eng 2007. 13:483–92. DOI: 10.1089/ten.2006.0181 121

[994] Ingber, DE, Mow, VC, et al. *Tissue engineering and developmental biology: going biomimetic.* Tissue Eng 2006. 12:3265–83. DOI: 10.1089/ten.2006.12.3265 123

[995] Jones, RE, Willie, BM, et al. *Analysis of 16 retrieved proximally cemented femoral stems.* J Arthroplasty 2005. 20:84–93. DOI: 10.1016/j.arth.2004.08.006 124

[996] Cadosch, D, Chan, E, et al. *Metal is not inert: role of metal ions released by biocorrosion in aseptic loosening–current concepts.* J Biomed Mater Res A 2009. 91:1252–62. DOI: 10.1002/jbm.a.32625

[997] Lemons, JE. *Retrieval and analysis of explanted and in situ implants including bone grafts.* Oral Maxillofac Surg Clin North Am 2010. 22:419–23, vii. DOI: 10.1016/j.coms.2010.06.002 124

[998] Dexter, TM, Allen, TD, et al. *Stimulation of differentiation and proliferation of haemopoietic cells in vitro.* J Cell Physiol 1973. 82:461–73. DOI: 10.1002/jcp.1040820315 124

[999] Greenberger, JS. *Sensitivity of corticosteroid-dependent insulin-resistant lipogenesis in marrow preadipocytes of obese-diabetic (db/db) mice.* Nature 1978. 275:752–4. DOI: 10.1038/275752a0 124

[1000] Zipori, D, Duksin, D, et al. *Cultured mouse marrow stromal cell lines. II. Distinct subtypes differing in morphology, collagen types, myelopoietic factors, and leukemic cell growth modulating activities.* J Cell Physiol 1985. 122:81–90. DOI: 10.1002/jcp.1041220113 124

[1001] *isolation of mononuclear cells from bone marrow aspirates by density gradient centrifugation.* Published by. Last Updated: 2008. Accessed on: December, 2010. Available at: http://www.miltenyibiotec.com/download/protocols_sample_preparation_en/1394/SP_MC_BM_density_gradient.pdf. 124

[1002] Wagner, W and Ho, AD. *Mesenchymal stem cell preparations–comparing apples and oranges.* Stem Cell Rev 2007. 3:239–48. DOI: 10.1007/s12015-007-9001-1 125

[1003] Cournil-Henrionnet, C, Huselstein, C, et al. *Phenotypic analysis of cell surface markers and gene expression of human mesenchymal stem cells and chondrocytes during monolayer expansion.* Biorheology 2008. 45:513–26. 125

[1004] Wagner, W, Wein, F, et al. *Comparative characteristics of mesenchymal stem cells from human bone marrow, adipose tissue, and umbilical cord blood.* Exp Hematol 2005. 33:1402–16. DOI: 10.1016/j.exphem.2005.07.003 125

[1005] Tolar, J, Nauta, AJ, et al. *Sarcoma derived from cultured mesenchymal stem cells.* Stem Cells 2007. 25:371–9. DOI: 10.1634/stemcells.2005-0620 125

[1006] Tasso, R, Augello, A, et al. *Development of sarcomas in mice implanted with mesenchymal stem cells seeded onto bioscaffolds.* Carcinogenesis 2009. 30:150–7. DOI: 10.1093/carcin/bgn234 125

[1007] Lazennec, G and Jorgensen, C. *Concise review: adult multipotent stromal cells and cancer: risk or benefit?* Stem Cells 2008. 26:1387–94. DOI: 10.1634/stemcells.2007-1006 125

[1008] Vogel, G. *Cell biology. To scientists' dismay, mixed-up cell lines strike again.* Science 2010. 329:1004. 125

[1009] Thomas, RJ, Hourd, PC, et al. *Application of process quality engineering techniques to improve the understanding of the in vitro processing of stem cells for therapeutic use.* J Biotechnol 2008. 136:148–55. DOI: 10.1016/j.jbiotec.2008.06.009 125

[1010] Steigman, SA, Armant, M, et al. *Preclinical regulatory validation of a 3-stage amniotic mesenchymal stem cell manufacturing protocol.* J Pediatr Surg 2008. 43:1164–9. DOI: 10.1016/j.jpedsurg.2008.02.052 125

[1011] Horwitz, EM, Le Blanc, K, et al. *Clarification of the nomenclature for MSC: The International Society for Cellular Therapy position statement.* Cytotherapy 2005. 7:393–5. DOI: 10.1080/14653240500319234 125

[1012] Dominici, M, Le Blanc, K, et al. *Minimal criteria for defining multipotent mesenchymal stromal cells. The International Society for Cellular Therapy position statement*. Cytotherapy 2006. 8:315–7. DOI: 10.1080/14653240600855905 125

Authors' Biographies

ERICKA M. BUENO

Ericka M. Bueno, Ph.D., obtained her doctorate in Chemical Engineering at Northeastern University (Boston, MA) in 2006 with a dissertation on bioreactors for cartilage tissue engineering. Her post-doctoral assignments involved cornea and cartilage tissue engineering, bone regeneration, and collagen biomaterials development. Dr. Bueno has authored several peer-reviewed articles in the field of tissue engineering and has contributed both oral and poster presentations at more than 20 international conferences. She is an active member of the Orthopedic Research Society. In 2008, she joined the Department of Orthopedic Surgery at Brigham and Women's Hospital and Harvard Medical School (Boston, MA) as a Post-doctoral Research Fellow. She is currently a researcher in composite tissue allotransplantation in the Division of Plastic Surgery at Brigham and Women's Hospital and, since 2010, the Manager of the Face and Hand Transplant Programs at Brigham and Women's Hospital, and Instructor of Surgery at Harvard Medical School.

JULIE GLOWACKI

Julie Glowacki, Ph.D., is Professor of Orthopedic Surgery, Harvard Medical School and Professor of Oral & Maxillofacial Surgery, Harvard School of Dental Medicine. She is Director of Skeletal Biology Laboratory at Brigham and Women's Hospital, Boston, MA and co-director of the Brigham and Women's Hospital Musculoskeletal Research Center. She received her Ph.D. in Biological Chemistry from Harvard University. She has published more than 130 original, peer-reviewed research articles and numerous chapters concerning basic, translational, and clinical research on skeletal development and aging, on the roles of vitamin D in bone formation and skeletal health, and on innovative approaches to bone and joint construction and reconstruction. She has developed a number of animal models and methods for defining effects of age, gender, and other clinical features on properties of human, marrow-derived, skeletal progenitor cells. She co-edited *The Aging Skeleton*. She has been a voting member of the US FDA Dental Products Panel and led their initiative on bone substitute materials. She has served on NIH, NASA, Department of Defense, international, and foundation grant review and advisory panels.

Printed in the United States
by Baker & Taylor Publisher Ser